W9-BUE-149

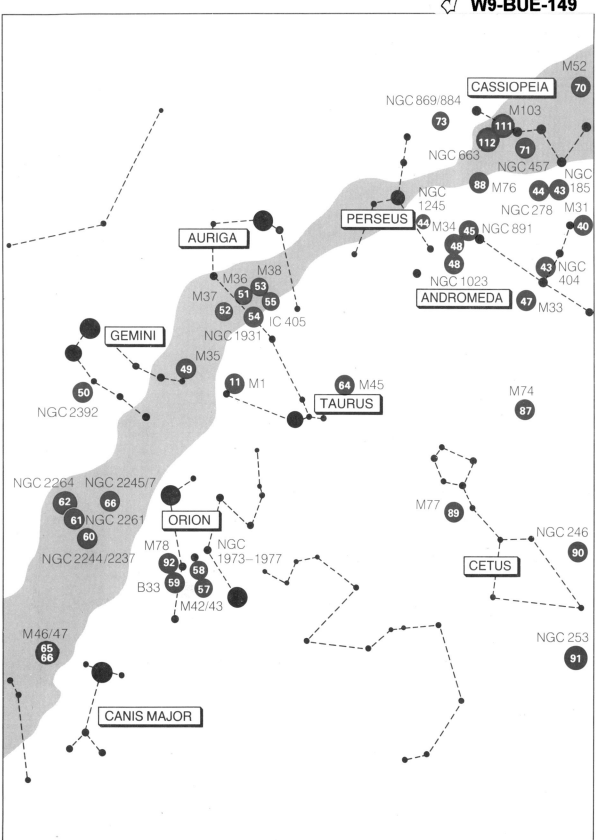

The Cambridge deep-sky album

Celestial photography by
Jack Newton

Text by
Philip Teece

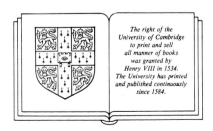

The right of the
University of Cambridge
to print and sell
all manner of books
was granted by
Henry VIII in 1534.
The University has printed
and published continuously
since 1584.

Cambridge University Press

Cambridge
London   New York   New Rochelle
Melbourne   Sydney

We dedicate this work to

Jack Newton's parents,
Dorothy and Merlin

And to our patient wives,
Janet and Wendy

Published by the Press Syndicate of the University of Cambridge
The Pitt Building, Trumpington Street, Cambridge CB2 1RP
32 East 57th Street, New York, NY 10022, USA
296 Beaconsfield Parade, Middle Park, Melbourne 3206, Australia

© Cambridge University Press 1983

First published 1983
Reprinted with corrections 1984

Printed in Great Britain

Library of Congress catalogue card number: 83-1430

British Library Cataloguing in Publication Data

Newton, J.
  The Cambridge deep-sky album.
  1. Astronomy – Observers' manuals
  I. Title   II. Teece, P.
  522   QB64

ISBN 0-521-25668-2

SE

CONTENTS

INTRODUCTION

Of all the branches of science, astronomy has always attracted the largest number of amateur and casual participants. This is perhaps because astronomy is the most visual of the sciences. The visible light of stars, nebulae and galaxies – historically the principal source of our scientific knowledge of the heavens – provides also the most obvious aesthetic pleasure to be found in any objects of scientific study. Since remote prehistoric times, man has loved to look up at the stars.

The present book has as its chief aim a presentation of the beautiful and intriguing visual appearance of objects in the night sky. That these celestial wonders are readily accessible to amateur observers is a certifiable fact; the authors themselves are amateur astronomers who have enjoyed these deep-sky objects for many years, with modest-sized telescopes.

The core of this presentation is an album of Jack Newton's colour portraits of the most spectacular (as well as many less well known) targets for ordinary amateur instruments. The photographs convey something of the lucid vitality of star clusters, luminous interstellar gas clouds and remote galaxies – a quality that makes the telescopic view so exciting, but which is never adequately conveyed in words. Jack Newton's technical introduction, immediately following this general foreword, describes the homebuilt telescope and cold camera by which the photographs were obtained.

The book is intended, however, to be much more than a gallery of celestial pictures. It is planned and arranged to serve as a practical observer's handbook for users of a wide range of amateur telescopes, from 5- to 50-cm aperture – and beyond, for those observers fortunate enough to have access to even larger instruments.

Some features of this handbook

For ease of reference, the 126 plates have been laid out in a numerical sequence, in order of the Messier numbers of the principal objects illustrated. Some important objects, however, do not have Messier numbers; they are identified only by designations from other catalogues (NGC, Barnard, etc.). Because these tend not to be well known by number, they are arranged in order of their positions near the better known fields containing Messier attractions. Thus, the observer who consults a particular region of the sky in search of a famous deep-sky object from the Charles Messier list may conveniently locate the less well known clusters, nebulae or galaxies that lie nearby.

Accompanying each plate depicting a photographic field, there is text that provides both practical observational comments and general scientific data on the chief objects within the field. It will be apparent that, although there are only 126 plates, many of them encompass several objects; thus, the total number of observational targets portrayed here is actually very large. The backgrounds against which the objects' headings are set are colour coded: red for galaxies, blue for nebulae, yellow for clusters.

Scale and orientation

Collections of celestial photographs, however beautiful and impressive, give only limited practical guidance if they are reproduced in a wide range of field-scales, and if the natural orientation of the north–south/east–west co-ordinates is not indicated.

All of the objects are here photographed on the same scale; the field in every case is a rectangle about one degree in length – the space occupied in the sky by two full moons. This uniformity is a crucial feature of the book, enabling one to gain an instant appreciation of the relative sizes of all the objects portrayed. The difference in size of the images of the globular clusters M13 and M56, for example, may be taken as a true indication of their comparative appearances through the telescope.

The colour plates are accompanied by small field-charts. In most cases these take the form of a simple compass-rose, indicating the approximate orientation of the field. Several of the diagrams, however, have been elaborated to include labelling of specific features that have been captured on the film.

Additional aids to observing

The locations of the chief deep-sky objects are shown in two ways: (A) The central feature of each photograph is identified at the head of the page by number and by co-ordinates for the telescope's setting circles. (B) These main, numbered features can be found on the charts of the sky-atlas, to be found on the front and back endpapers.

A small-telescope astronomer who wishes to form some estimate of an object's brightness will be guided by the item labelled 'Mag' in the panel of data at the top of the text-page; this is the *apparent visual magnitude.** Very roughly, it may be said that galaxies and nebulae in the range of 4th to 8th magnitude will be most suitable for modest instruments of 7- to 10-cm aperture. Those in the magnitude range between 8 and 10 will begin to be interesting only if viewed with something larger – a 15- or 20-cm telescope. Really impressive views of the fainter objects (below 10th magnitude) require a mirror with substantial light-grasp, in the 30- to 50-cm class.

---

* For consistency, all magnitudes have been quoted from the same source – the tables of data to be found in *The Observer's Handbook* of the Royal Astronomical Society of Canada.

Tables of data

For quick reference, the reader will find a compendium of basic data in tabular form at the end of the book. The tables include comparative information on not only the major objects of the plates, but also on several fainter and less familiar galaxies, etc., that appear in the photographs.

It is sometimes a fault of catalogues and lists for the amateur observer that no distinction is made between bright, easy sky-targets and those that may be expected to be a much more difficult challenge. (In the case of extended objects, the figures quoted for visual magnitude are often relatively meaningless.) A particularly useful feature of the tables, therefore, is the column that gives a rough indication of the minimum size of telescope required for a satisfactory view of each object.

Special terms and abbreviations

Catalogue-designations:

M   Messier (Charles-Joseph Messier)
NGC   New General Catalogue (John Dreyer)
IC   Index Catalogue (Supplement to NGC)
B   Barnard (Edward E. Barnard)
A   Abell, George O.
Mel   Melcotte Catalogue (an early list of nonstellar objects)

Chart-designations and positional information:

RA   Right Ascension; the west–east position on the sky, given in hours (h) and minutes (m) eastward from the First Point of Aries
Dec   Declination; the vertical co-ordinate in the sky, given in degrees (°) and minutes (′) above or below the Celestial Equator
Greek letter designations of stars: The brightest stars in each constellation are identified by letters of the Greek alphabet, usually in descending order of brightness (i.e. 'Alpha' is the first star of the constellation, 'Beta' the second brightest, etc.)

The Greek alphabet

α	Alpha	ι	Iota	ρ	Rho
β	Beta	κ	Kappa	$\sigma\,(\Sigma)$	Sigma
γ	Gamma	λ	Lambda	τ	Tau
δ	Delta	μ	Mu	υ	Upsilon
ε	Epsilon	ν	Nu	φ	Phi
ζ	Zeta	ξ	Xi	χ	Chi
η	Eta	o	Omicron	ψ	Psi
θ	Theta	π	Pi	ω	Omega

Miscellaneous abbreviations:

E.G.   Elliptical galaxy
G.C.   Globular cluster
Irr.   Irregular
Mag   Magnitude
O-, B-, A-, F-, G-, K-, M-class   Letters indicating the principal types of stars, the chief differentiating characteristics being temperature and colour (O, B, A – very hot, white; F, G, K – moderate, yellow; M – cool, red)
O.C.   Open cluster
Parsec   A standard unit of interstellar distance = 3.26 light-years
Sa (or Sb or Sc) galaxy   Spiral galaxy. The lower-case letters indicate three sub-classes of spirals whose arms are tightly (Sa), moderately (Sb) or loosely (Sc) coiled about the galactic core

A note on the term 'One Billion'

This term, which has traditionally been understood differently on opposite sides of the Atlantic, is here in every instance used to represent the figure one thousand million (10^9).

TELESCOPE AND CAMERA

Most of the photographs in this *Deep-Sky Album* were taken with a homebuilt 16-inch (40-cm) reflecting telescope, whose 2000-mm focal length is a relatively 'fast' f/5. This instrument is housed in a 10-foot (3 m) domed observatory atop a small hill, eight miles (13 km) west of Victoria, British Columbia.

The camera also is a homemade product. It is a cold camera designed by Jack Newton and constructed of inexpensive plastic by another Victoria amateur, George Ball. In fact, the latter's rôle was to listen to Jack Newton's verbal description of a mechanical or optical idea, and then to design and build a mechanism that met the indicated need. The principle of the resulting celestial camera, and the technique of using it in conjunction with the astronomical telescope, will be described below, in some detail.

The photographs were taken on 35 mm Ektachrome 400 slide film, exposed for about fifteen minutes. These exposures record all stars brighter than 18th magnitude.

The telescope

The chief optical element of this Newtonian reflector is its 40-cm diameter main mirror. The large disc of Pyrex glass from which the mirror evolved was owned for thirty years by Mr Al Donnelly, who planned someday to use it in one of his telescope building projects. The fine old mirror-blank's long term of cloistered existence ended in 1980, when Jack Newton persuaded its owner to part with it; reluctantly, Mr Donnelly agreed to its sale.

With the assistance of Jack Newton, Leo Van derByl immediately set to work on the grinding and polishing of the 25 kg glass blank, rapidly generating the required f/5 curvature and polishing the surface to an optical figure of the highest precision. Thus, in just thirty days the blank that had lain idle for thirty years became a finished astronomical mirror.

The mirror is mounted inside a steel tube of 43 cm diameter; the walls of this tube act as the supporting cell for the mirror, which rests on three Teflon pads, and is centred by Teflon wedges that slide between the edges of the mirror and the inner surfaces of the tube. At the upper end of the instrument, a diagonal mirror (a small elliptical flat of 9 cm minor axis) deflects the light to the guiding head and camera, mounted on the side of the telescope.

Jack Newton's domed observatory

At the time of writing, this dome has just been replaced by a large semi-domed structure that will soon house a new 50-cm telescope.

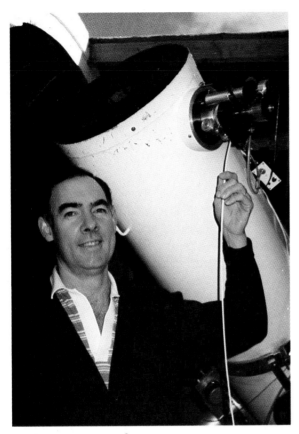

A 40-cm amateur telescope

With this large homebuilt instrument, mounted on a hilltop near Victoria, British Columbia, Jack Newton photographed all the celestial objects portrayed in this book. (All are fourteen-minute exposures with a cold camera that chills the film for optimum results.)

In long-exposure celestial photography, good optics are not even half the secret of success; at least equally essential is a firm and perfectly precise means of supporting and moving the telescope. In the case of the 40-cm reflector, a massive equatorial mounting, weighing over 225 kg, carries the telescope, and a 25-cm Mathis worm gear powered by a Hurst motor provides the drive, to counteract the effect of the earth's rotation during the exposure. The speed (number of revolutions per minute) of the motor can be altered by a variable frequency control, activated by 'fast' and 'slow' push-buttons. Precise adjustments of the telescope's position on the declination axis are made by a reversible motor with a centre-off switch.

Guiding the telescope

A common technique for guiding during long exposure involves the use of a smaller telescope mounted on the tube of the main instrument. The astronomer monitors the accuracy of the driving motion, during a prolonged exposure, by keeping a selected star centred on the guide-telescope's crosshairs.

In this standard system, however, photographic precision can be marred by flexure that occurs between the guide-telescope and the instrument on which the camera is mounted. The result of this variance between the two optical paths is a photograph on which the star images are not small and circular, but elongated or 'trailed'. The guide-telescope method poses additional difficulties: the small refractor's modest light-grasp, relatively narrow field and restricted mobility greatly limit the selection of stars suitably bright and well placed for guiding.

A more satisfactory system is used on the 40-cm Newtonian. This telescope's 'off-axis guiding head' consists of a 12-mm eyepiece and Barlow lens combination (magnification: 500 ×) mounted through the main instrument's rotatable focusing mechanism. A set of pivoting prisms direct light from the 40-cm primary mirror itself into the guiding eyepiece; by moving these prisms, and rotating the 'head', one may select a guide-star at any location within the field. Because the same optical system is supplying both the set of images that are to be photographed and the star image by which the photograph will be guided, there is no discrepancy of motion between the two. Crisp, round, untrailed impressions of the stars on the developed film are the end result.

The cold camera

The photographs that form the core of the present book were all taken with a homemade camera designed to use 35 mm film for cold temperature photography.

Most films, when exposed for more than a few seconds, suffer from reciprocity failure – a dramatic retardation of photochemical activity in the emulsion. Thus, the reciprocal relationship of image-intensity and time ceases to hold true; a film exposed twice as long fails to accumulate twice the density of image.

The formation of a picture on exposed film is caused by a chemical change in the crystals of the photographic emulsion, triggered by incident light. During long exposures at low light levels, the accumulation of light may be so slow that many of the activated crystals are already decaying while others are still being induced to react. The cold camera is a solution to this problem.

In the cold camera, the crystals of the film-emulsion are very deeply chilled, to a temperature at which they hold the accumulating image longer. In this state, the decay of activated crystals is arrested; a slow buildup of light is thus able to bring about on the film a greater net chemical change that will appear, during development, as a strong permanent image.

The accompanying drawings illustrate two versions of the cold camera. Both consist of three basic sections: a shutter and spacing-tube (which plugs into the telescope), a camera body that holds the film, and a dry-ice container at the back of the camera. The simpler design uses a solid optical-plastic plug to fill the airspace between film and shutter. In the more sophisticated model, this plug is replaced by a chamber filled with dry nitrogen gas. One end of this chamber lies in contact with the film; at the other end (adjacent to the shutter) a heated glass window keeps the chamber sealed. The purpose of either the optical plug or, alternatively, the nitrogen-chamber is to prevent moisture and frost from appearing on the chilled film during the long exposure.

The photographic routine

A typical photographic run begins with the shutter portion of the camera attached to the telescope, and with an eyepiece placed on the dry-gas chamber. The celestial object to be photographed is centred in the field of view observed through the eyepiece, and then a suitable guide-star is located with the guiding head. Next, the camera is focused, using a sharp edge at the film plane. The shutter is closed and, when the camera body has been placed over the dry-gas chamber, nitrogen gas is flushed through the chamber. Crushed dry ice, taken now from the Thermos bottle in which it is stored, is placed into the camera. Finally, the shutter is opened and the exposure is made.

On completion of the fifteen-minute exposure the shutter is closed, and the camera body is carefully removed and capped. (Excess dry ice is returned to the Thermos bottle.) After a hairdryer has been focused on the camera-back for four minutes to raise the film to ambient temperature, the film is advanced, ready for the next exposure. The whole cycle takes about one hour.

An alternative approach

'Gas-hyperization' of film has recently become popular with amateur astrophotographers. This is a method of reducing the effects of reciprocity failure without having to use the cold camera at all. Film that has been sensitized by pressured flushings of a noncombustible forming gas (a mixture of 92% nitrogen and 8% hydrogen) can be used in a standard camera, with dramatic results.

The amateur photographer can make a pressurized tank for this purpose from black plastic plumber's pipe, available in most hardware stores. A welding supply shop can provide the necessary pressure-gauge, and an old bicycle tube is a source of an air-valve. The gauge is cemented into one end of the pipe and the tyre-valve into the other end. The entire project can be completed in less than an hour, for about the price of a bottle of Scotch.

The film is often 'baked' during the hyperization process. The baking chamber can be constructed from a Styrofoam freezer equipped with a 60-watt light bulb and a thermostat.

The procedure for sensitizing film is as follows: After the film-cassettes have been flushed three times with forming gas in the hyperizing tank, the tank is pressurized to 15 pounds per square inch (1 kg per cm²). The tank is placed in the baking chamber for four days at 40 °C. Once each day the hyperizing tank is flushed. Film-cassettes treated in this fashion are sealed into taped plastic canisters and stored in a freezer until needed. Since maximum storage time is about thirty days, it is sensible to hyperize only small amounts of film, to be used immediately.

Best results have been achieved with Kodak 2415 Technical Pan black-and-white, an extremely fine-grained film whose ASA rating approaches 400 after gas treatment. Ektachrome 400 has also proved moderately successful. It should be noted that the cold camera outperforms hyperized colour film. Treated Kodak 2415 black-and-white, however, yields superb results.

The new kodacolor VR1000 print film in combination with the Lumicon deep-sky filter (which replaces the glass window in the cold camera) has yielded stunning results. See pages 33, 36 and 37 for examples.

Darkroom procedure

Darkroom procedure is almost as important as the original photography itself.

Because most amateurs' telescopes are located near populous, light-polluted areas, exposure time is usually governed by sky-fog, rather than by limiting magnitude. At Jack Newton's location, for instance, colour Ektachrome 400 exposed for twenty minutes at f/5 badly overexposes the background sky-fog, on even the darkest nights.

The low contrast of faint objects can be somewhat enhanced by copying the original slide on Ektachrome 64. The colour-balance is often shifted slightly from pale blue-green to a dark blue-black, which is more aesthetically desirable.

The overexposed night sky is laced with mercury-vapour light-pollution, and is green-tinted. Pale aurora adds to this problem. Copying partially corrects these faults. Jack Newton uses a simple slide-copier that attaches to a 35 mm camera, much like a telephoto lens. A small electronic flash is used as a light source. A proper exposure is achieved only after some experimentation to determine the optimum distance of this source. Using the same procedure, one can produce colour negatives from original slides.

Small tracking errors are another fault that can be remedied in the darkroom. By stacking two negatives together in such a fashion as to offset the images on one, relative to the other, one may produce increased contrast and rounder star images. This stacking method also averages grain in the film, to produce a smoother background.

Cold camera I

This is the simpler of two cameras for celestial photography designed and built by Jack Newton and George Ball. Both versions use dry ice to chill 35 mm roll film, in order to increase the sensitivity of the emulsion during long exposures. In this version (ideal for the beginner at astrophotography), a simple clear-plastic plug fills the airspace between shutter and film.

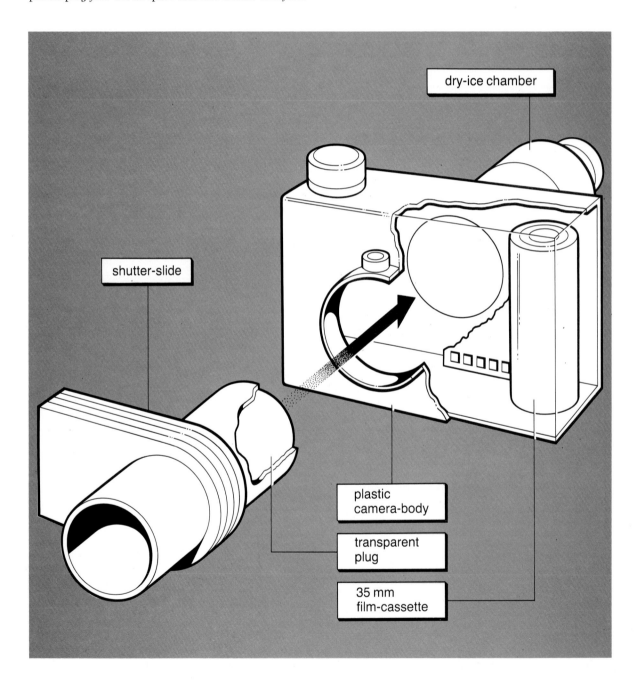

dry-ice chamber

shutter-slide

plastic camera-body

transparent plug

35 mm film-cassette

Cold camera II

Camera body (not shown) fits over chamber, at left side of drawing. In this more sophisticated arrangement, dry nitrogen gas fills the space between shutter and film.

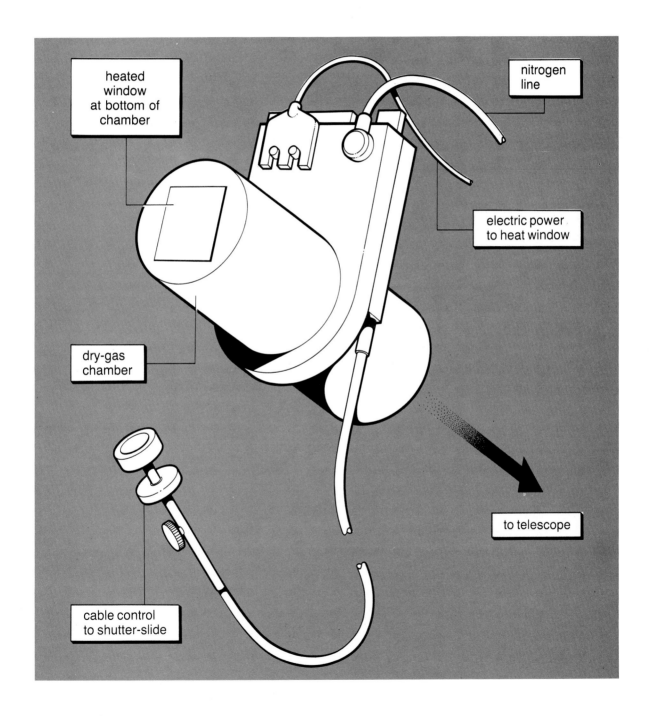

heated window at bottom of chamber

nitrogen line

electric power to heat window

dry-gas chamber

to telescope

cable control to shutter-slide

THE ALBUM OF DEEP-SKY OBJECTS

M1 The Crab Nebula
NGC 1952
Supernova remnant in Taurus

RA   05h 33m

Dec +22° 01′

Mag   8.4

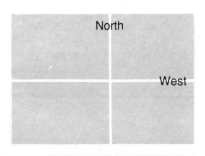

North

West

In July of the year AD 1054, there appeared a dramatic celestial portent – a star that brightened to such prominence that it could be observed even in the daylight hours. Seven centuries later, at the same position in the sky (about one degree northwest of the star *Zeta* in Taurus), Charles Messier's small telescope showed a strange, nebulous smudge.

Messier-1, the Crab Nebula, is now regarded as quite certainly the splash of debris left over from the supernova explosion observed in 1054. Because its intriguing identity, and also because of many puzzling features that characterize it, the Crab has been the subject of unusually intense study by astrophysicists.

Early in the present century, it was seen that photographs of the nebula taken over a significantly large interval of time actually revealed changes in the Crab's shape and size; 900 years after the initial explosion, the resultant envelope of gas was still visibly expanding.

In 1968 the nebula's famous central pulsar was discovered. This faint star embedded in the heart of the Crab emits continual bursts of radio energy at the almost unbelievable frequency of thirty pulses every second. Clearly, it is a small, compact object with an extraordinary rate of spin – a neutron star, the superdense object left behind as the remnant 'cinder' of the original supernova.

The effects of this star's energetic behaviour within its envelope of ejected gas are varied and peculiar. The nebula's X-rays and polarized light suggest that it behaves like a naturally occurring synchrotron. One theorist has suggested that multiply-reflected optical emissions from the nebula's core make the Crab Nebula a sort of cosmic laser.

M2
(NGC 7089)

RA   21h 32m
Dec −00° 54′

This 6th-magnitude globular cluster in Aquarius is about 50 000 light-years distant. A 20-cm telescope will resolve some of its several hundred thousand individual stars.

NGC 7814

RA   00h 01m
Dec +15° 51′
Mag   12.0

About 40 degrees northeast of M2, in Pegasus, this 12th-magnitude spiral galaxy is viewed edgewise, displaying an equatorial lane of obscuring dust. While the globular M2 may seem a remote object, NGC 7814 is vastly more distant – tens of millions of light-years from us.

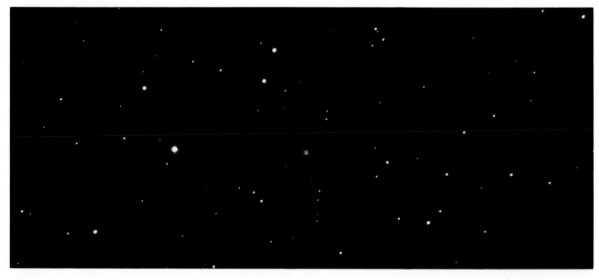

M3
NGC 5272
Globular cluster in Canes Venatici
RA   13h 41m
Dec +28° 29"
Mag   6.4

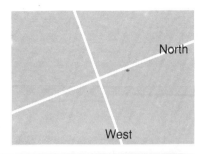

If you sweep westward from the middle regions of the constellation Bootes, using good binoculars, you will stumble upon a very faint circular patch of light. The 40-cm telescope's view of that same object, reveals the true nature of M3: it is a very distant, but enormous, globular star-system. Probably as many as half a million stars crowd together within its 200-light-year diameter. This massive cluster is no near neighbour; it is of the order of 40 000 light-years away from us.

Early in the present century Harlow Shapley discovered the significant relationship of the globular clusters to the Milky Way spiral system, in which our Sun is located. The hundred or so of these objects that we observe form, as Shapley learned, a roughly spherical halo lying above and below the plane of our galaxy.

Shapley recognized the usefulness of this arrangement of the globulars: since they lay well outside the obscuring dust of the galactic plane, there was nothing to hinder measurement of distances to these clusters – even those that lay in remote regions beyond the centre of the galaxy. Because the halo of globulars appeared to be centred on the galactic centre, the scale of the globular swarm provided the first useful estimate of our distance from the galactic core. Thus, the globulars were the tools by which we came to realize the size of our galaxy (about 100 000 light-years' diameter).

The astronomer's key to measurement of great distances is the fixed relationship of period to luminosity in a certain class of stars, the 'Cepheid' variables. The length of time required for these stars to perform their characteristic variation of brightness is proportional to their true luminosity. Comparison of the estimated true luminosity with the star's brightness as observed from the earth yields a measure of distance. M3 is a 'pioneer' globular, for it was the first in which a periodic variable star was discovered.

M4
NGC 6121
Globular cluster in Scorpius
RA   16h 22m

Dec $-26°$ 27′

Mag   6.4

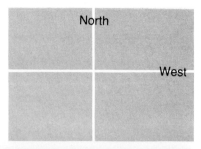

The balefully glaring red star Antares, in the abdomen of the Scorpion, is one of the summer sky's most notable signposts. Only about a degree west of this famous luminary is the fine globular cluster illustrated here.

Although, at visual magnitude 6.4, M4 should be as prominent as the more celebrated Hercules cluster, it tends to disappoint observers at relatively high northern latitudes. This diminution of its potential lustre is due to the obscuring haze through which an object low above the southern horizon is usually viewed. Nevertheless, if conditions are favourable, a 15-cm telescope will resolve some of the individual stars in the globular's outer regions.

Compared with most of the well known globular clusters, M4 is an unusually nearby object; at only about 14 000 light-years, it is much less than half as remote as M3. (For an extreme comparison, see NGC 7006, a 'maverick' globular cluster.)

M4 is a large cluster, filling about the same portion of a telescopic field as the great Hercules cluster (M13). Being relatively loose, however, and without the typically dense central condensation, M4 appears to be a considerably less massive system than M13, with fewer members.

M5
NGC 5904
Globular cluster in Serpens
RA   15h 18m
Dec +02° 10″
Mag   6.2

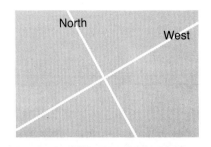

Lying at a distance roughly comparable to that of the great Hercules cluster (M13), this superb globular is only slightly less impressive when viewed with the telescope. To gain an appreciation of the effect of distance on the appearance of these objects, it is instructive to compare relatively nearer clusters such as M5 and M13 with the more remote globulars M56 and NGC 7006. (The photographs in this book are all on the same scale.)

M5 is a major globular cluster; like the Canes Venatici cluster described on page 13 and the great Hercules cluster,

it is a massive and highly compact aggregation of many hundreds of thousands of stars.

The amateur telescopist will find M5 intriguing for several reasons. Having numerous members as bright as magnitudes 11 and 12, the cluster quite readily shows some resolution into individual stars when viewed with a good 15-cm telescope. At least two of this globular's variable stars are unusually prominent (approaching 10th magnitude at maximum), and conveniently located in the sparse fringe-regions where they can easily be identified. An amateur equipped

with an instrument of moderate aperture will find it interesting to monitor the $1\frac{1}{2}$-magnitude variations of stars 42 and 84.*

Keen-eyed observers looking at M5 for the first time usually notice a delicately attractive double star that lies in the same field, at low magnification. It will be noticed that 5 Serpentis, near the globular's southern fringes, has a close, faint companion.

* C.M. Coutts Clement and Helen S. Hogg have given a detailed description of these stars in: 'The bright variable stars in M5', *Journal of the Royal Astronomical Society of Canada*, Aug. 1977, pp. 281–97.

M8 The Lagoon Nebula
NGC 6523
Diffuse nebula in Sagittarius
RA   18h 02m

Dec −24° 23′

Mag   6.0

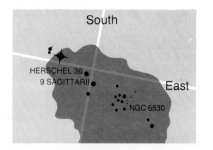

On a clear, moonless summer night, if one looks at a point in the southern sky about 6 degrees north of the star γ *Sagittarii*, one may detect even with the unaided eye the dim, unfocused glow of the Lagoon Nebula. A small telescope reveals much of this magnificent object's splendour: its sparkling star cluster (NGC 6530), its pale billows of nebulosity and the dark void of the obscuring dust lane across its centre.

M8, one of the greatest of the H II (ionized hydrogen gas) regions in our galactic neighbourhood, is a complex object. The star cluster appears to heat a dense molecular cloud that lies behind it. The very compact, black 'globules' that lie in the foreground, silhouetted against the glow of the H II region, are believed to be cosmic material far advanced in the gravitational collapse that gives birth to new stars. NGC 6530 itself is believed to be a cluster of some of the youngest stars known; probably as recently as two million years ago they were, themselves, dark globules that had not yet become luminous.

The two brightest field-stars, prominent in the telescopic view of M8, are 9 Sagittarii and Herschel 36. These are not relatively nearby foreground objects, but highly energetic stars associated with the nebula itself; they have been identified as the principal sources of lumination of the surrounding nebulosity.

The Lagoon's distance is a matter of some uncertainty. The Royal Astronomical Society of Canada's *Observer's Handbook* cites a distance of 4500 light-years. *Burnham's Celestial Handbook* accepts the evidence for a much remoter position, about 5200 light-years away from our solar vicinity. If the latter estimate is used, the Lagoon is an enormous interstellar cloud, about 100 light-years wide. On this scale, the full diameter of our solar system is one hundredth that of the smallest dark globules observed in the Lagoon.

M10
NGC 6254

RA   16h 56m
Dec −04° 05′
Mag   6.7

Easily located with binoculars, this globular cluster in Ophiuchus is about 20 000 light-years distant. A good 15-cm telescope will just begin to resolve some of its individual stars.

M12
NGC 6218

RA   16h 46m
Dec −01° 55′
Mag   6.6

Only about $3\frac{1}{2}$ degrees west of M10, M12 is a somewhat fainter globular, but slightly easier to resolve because of its looser and more sparse stellar membership. Both clusters are about equally distant from us.

M11 The Wild Duck Cluster
NGC 6705
Open cluster in Scutum
RA 18h 50m
Dec −06° 18′
Mag 6.3

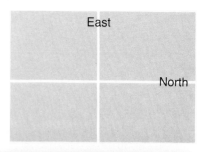

East

North

Most descriptive handbooks agree in their assessment of M11 as probably the finest open cluster. It was the early 19th-century amateur, Smythe, who noted the V-formation in M11 that suggests the simile of wild ducks in flight, and his contemporary, T.W. Webb, described the object as 'a noble fan-shaped cluster'. A modern 15-cm telescope gives the same impression of this group.

Even relatively brief photographic exposures reveal the rich population of M11 – a dense crush of probably 1000 stars within a quarter-degree circle. With average interstellar separations of only about one light-year in the core-region, this galactic star-swarm is remarkably like a miniature globular cluster.

At the cluster's distance of 5600 light-years, ordinary dwarf stars like our Sun are virtually imperceptible; most of the cluster members captured on the plate are giant stars of very high luminosity. To attain a full realization of M11's compactness and density we must, in imagination, fill the spaces with several hundred fainter stars of sun-like mass and brightness.

On a hazy or moonlit night, a newcomer to astronomy may find the Wild Duck Cluster disappointingly featureless (17-century astronomers, with their primitive small refractors, saw it only as a dim patch of mist). In suitable conditions, however, even a little 6 cm telescope will cause this nebulous spot to dissolve into a breathtaking spray of minute, dust-like stars.

NGC 6781
Planetary nebula in Aquila
RA 19h 16m
Dec −06° 26′
Mag 12.0

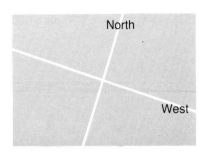

North

West

There is a kind of minor catastrophe that appears to be not infrequent in our galaxy. Typically, during the late stages of a star's evolution, it may undergo some degree of sudden readjustment (not so violent as the explosion of a supernova), resulting in the ejection of material into surrounding space. The compact, roughly spherical shell of gas that envelopes such a star has been described as a 'planetary' nebula. Perhaps the best known object of this class is the Ring Nebula in Lyra (M57).

This ring nebula in Aquila is of slightly larger apparent size than M57, but 3 magnitudes fainter. Of the several stars that can be resolved within the periphery of the gaseous shell, one – a 15th-magnitude dwarf – has been identified as the collapsed remnant of the star that ejected the material that forms the surrounding envelope.

Like the majority of planetary nebulae, NGC 6781 lies at a distance from us of some thousands of light-years, the precise figure being uncertain because of the difficulty of ascertaining the intrinsic brightness of the embedded dwarf star.

For the visual observer, this neat little planetary is a subtle target, requiring a dark, transparent sky and at least a moderate aperture. Contributors to the *Webb Society Deep-Sky Observer's Handbook* compare its appearance, as viewed with a 20-cm telescope, with that of the rather elusive Owl Nebula (M97). Only through a considerably larger instrument does the vague disc of NGC 6781 clearly resolve itself into its annular form.

The Saturn Nebula
NGC 7009
Planetary nebula in Aquarius

RA   21h 01m

Dec − 11° 34′

Mag   9.0

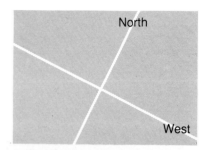

North

West

NGC 7009 is a prominent example of the class of objects to which Herschel and other early astronomers applied the name 'planetary' nebulae. The telescopic appearance, typically a very small but well defined disc, seemed analogous to that of a distant planet.

Of NGC 7009 we find in T.W. Webb's 19th-century notebooks a description that still gives a valid impression of the nebula as viewed with an amateur's instrument. With his fine old 5-inch (13-cm) Alvan Clark refractor Webb saw this planetary as 'somewhat elliptic: very bright for an object of this nature; pale blue; not well defined . . . but bearing magnifying more like a planet than a common nebula'.

The nebula is a small bubble of gas surrounding a central star that lies some 3000 light-years distant from us. Its overall diameter is about half a light-year. In the photograph we are able to see the atypical feature of the Saturn Nebula that prompted Rosse, in 1850, to coin that name: the ovoid disc is flanked by two narrow projections of luminous material that bear a resemblance to Saturn's rings, when they are viewed edgewise. The traditional interpretation of these appendages is that they are in fact a flattened ring of gas whose plane lies approximately on our line of sight.

The Helix Nebula
NGC 7293
Planetary nebula in Aquarius
RA   22h 27m

Dec −21° 06′

Mag   6.5

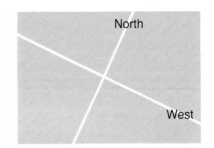

North

West

This very large planetary nebula, which appears as a featureless disc when observed visually, emerges on the photographic emulsion as a ring nebula similar to the better known M57 in Lyra.

Like a number of similar objects described in the present book (cf. M27, NGC 7662, M57), this vast gaseous 'ring' is actually a roughly spherical shell of material ejected from the faint star at its centre. As is typical in the planetary nebulae, the central star is a dense and hot, but exceedingly small, body – a dwarf star perhaps only twice the diameter of the earth. From the size and rate of expansion of the nebula's outer perimeter, it can be estimated that the catastrophic phase of the central star's evolution that hurled the material outward must have occurred a few tens of thousands of years ago.

The large apparent size of the Helix Nebula (compare it, for instance, with NGC 7009 on the previous page) is an effect of its location in space. The Helix is the most nearby of all known planetaries; its fairly average ring-diameter of approximately $1\frac{1}{2}$ light-years is made to appear unusually impressive because of its distance of only about 300 or 400 light-years.

Because of its broad extent and rather low surface brightness, this object is best observed with the telescope at lowest magnification. Large-aperture binoculars show the dim, sprawling planetary well, in the context of a wide star-field.

M13 The Great Hercules Cluster
NGC 6205
Globular cluster in Hercules

RA   16h 41m

Dec +36° 30′

Mag   5.7

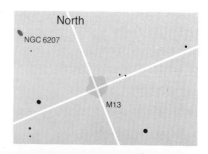

Of all the globulars visible to observers in northern latitudes, the great Hercules cluster is the most spectacular. Even as small an instrument as a 10-cm telescope shows some resolution of the peripheral regions into individual stars.

To its earliest watchers (Halley in 1715 and Messier and Bode in the 1770s), M13 was a mere patch of light. William Herschel first resolved this monstrous star-swarm, and recent studies have shown the staggering reality of the object: it is a dense crush of probably one million stars within a spherical space of somewhat over 100 light-years diameter.

Like the other globular clusters attendant upon our galaxy, M13 is a massive stellar system in its own right, and yet its 25 000-light-year distance reduces its luminous blaze to a mere 6th-magnitude smudge in our summer sky; an acute observer may just barely perceive it with the unaided eye. The faint (11th-magnitude) stars that one resolves with the telescope are the cluster's brightest members, red giants whose individual luminosities are equivalent to 2000 suns.

Visually, the cluster appears to spread outward from its centre to a more or less well defined boundary at about 6 arc-minutes radius; 90 per cent of M13's mass lies within this sphere. A recent proper-motion survey, however, done with the 100-cm Clark refractor at Yerkes (*Astronomical Journal*, vol. 84, pp. 774–82, 1979), has identified hundreds of remoter members lying as much as 12 arc-minutes from the centre of the cluster. Thus, the overall diameter of M13 is probably of the order of 170 light-years.

About 40 arc-minutes to the northeast of M13, an 11th-magnitude galaxy (NGC 6207) is visible.

A 2151

Cluster of galaxies in Hercules

RA 16h 05m

Dec +17° 40′

Mags approx. 15

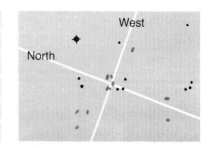

This highly concentrated aggregation of galaxies is a small part of the vast Hercules Supercluster – actually a cluster of galaxy-clusters!

The group shown in the photograph (Abell 2151) has been called 'the classical Hercules cluster'. The brightest members, of approximately visual magnitude 15, have been found to have redshifts indicative of velocities on the order of 10 000 kilometres per second. Using the Hubble correlation between retreat-velocities and distances, one derives a probable distance of about 600 million light-years for this remote family of star systems.

Extremely close, disruptive encounters between galaxies (a dramatic example is seen in the case of the two large, face-on spirals here) seem to be relatively commonplace within clusters. The tidal interaction between such pairs probably results in a loss of mass from both galaxies, in the form of gas 'spilled' into the intergalactic void. The clusters of the Hercules region exist within a tenuous envelope of such material, which astronomers describe as the 'intracluster medium'.

The giant radio-telescope at Arecibo in Puerto Rico was used recently to study the effects of the intracluster medium on individual galaxies within the Hercules clusters (Giovanelli *et al.*, *Astrophysical Journal*, vol. 247, pp. 283–402, 1981). These galaxies appear to be 'swept' by the envelope of intergalactic gas through which they move; thus, being stripped of neutral hydrogen gas, they are typically much freer of interstellar H I than, for instance, our Milky Way system.

M15

NGC 7078

Globular cluster in Pegasus

RA   21h 29m

Dec +12° 05′

Mag   6.0

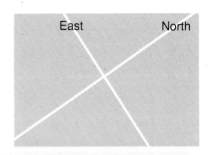

It is interesting, after carefully studying the great Hercules cluster with a telescope, to turn the instrument next toward M15 for comparison. Sixty per cent further away than M13, the Pegasus cluster (39 000 light-years distant) yields much less easily to resolution by a 10- to 15-cm aperture.

Even a small telescope shows, however, that these two fine globulars are dissimilar in structure. M15 is characterized by an unusually small, bright central nucleus, surrounded by outer regions that are relatively dim and scattered. We are looking at one of the most condensed of all the globulars, and one that is characterized by its own special hallmarks.

The core embodies a greater than normal concentration of red giant stars, compared with similar globulars such as M3, M5 and M13. It is also a fairly prominent X-ray source; some theorists have suggested the gravitational field of a massive 'black hole' within the cluster as a possible explanation of this radiation.

The Pegasus cluster is the only globular in which another feature has been observed. When an exceedingly tiny planetary nebula, of only about 14th magnitude, was photographed in 1927, it was at first thought to be possibly a foreground object lying relatively nearby on our line of sight. It is now believed that this shell of gas surrounds a star actually situated within M15.

For the casual astronomer in the northern hemisphere, M15 is one of the half-dozen most impressive globular clusters. A common 6-cm refractor shows it as a bright glowing mass; in a 20-cm telescope, it becomes a blazing multitude of tiny stars.

NGC 7331
Spiral galaxy in Pegasus
RA   22h 35m

Dec +34° 10′

Mag   9.7

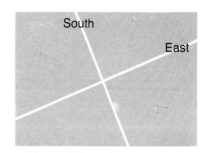

About 20 degrees northeast of the Pegasus globular cluster described on the previous page, NGC 7331 can be seen visually with a 15- to 20-cm telescope as a pale oval smudge of light. The object emerges on film as a galaxy viewed at an oblique angle similar to that of the better known M31 in Andromeda.

Like the Andromeda system, NGC 7331 is a spiral galaxy. Although its diameter and mass appear to be somewhat smaller than those of our own Milky Way or M31, it is a galaxy structured on the same plan, with a dense nucleus and moderately loose arms whose curving lines are marked by absorbing dust lanes and chains

of luminous nebulae. It is a star-system of more than 100 billion suns, situated at a distance of about 45 million light-years.

The Pegasus galaxy provides significant evidence for the debate about the direction of rotation of galactic spiral arms. The key to this question is the obscuring dust lane that lies along the spiral's western rim. In the best available photograph (see plate 17 in Allan Sandage's *Hubble Atlas of Galaxies*) this black lane is shown clearly as a foreground feature cutting across the luminous nucleus behind. With the western part of NGC 7331 thus identified as the edge of the galaxy nearer to us, it can be said that this galaxy rotates

in a direction such that the arms are trailing. The point at which a spiral arm joins the galactic core is the arm's leading edge, in rotation.*

Of the other galaxies that can be seen in the field, most are distant background objects. Three of these objects (adjacent, to the west) are thought to be associated with NGC 7331 itself. This apparent connection raises a difficulty that is related to the conundrum of the strange galaxy-cluster discussed on the following page.

* Like many solutions to astronomical problems, this conclusion does not meet with unanimous acceptance.

Stephan's Quintet
NGC 7317, etc.
Galaxy-cluster in Pegasus
RA 22h 33m
Dec +33° 42′
Mag 15.0

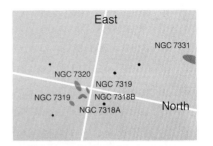

East

NGC 7331

NGC 7320

NGC 7319

NGC 7319   NGC 7318B

NGC 7318A   North

The quintet of rather distant galaxies, NGC 7317, 7318A, 7318B, 7319 and 7320, is an enigma that has been the subject of considerable speculation and controversy.

There are two strong indications that all five are physically associated in space as a tightly knit cluster. The resolvable H II regions in all of these galaxies have approximately the same apparent size – evidence of equidistance. Also, long exposures taken with the largest telescopes reveal a tenuous network of intergalactic material that seems to link the five galaxies.

The puzzling feature is this: while the light received from four of the group exhibits a redshift suggestive of a retreating velocity in the range of 5700 to 6700 km/sec, one member (NGC 7320) shows a velocity of only 800 km/sec. Some theorists have suggested that, if the five galaxies are really together in space, the situation casts doubt on cosmologists' use of retreating velocity as an indication of distance. The most plausible answer is that the cluster itself is flying apart; the high-speed dispersal of the member-galaxies (some away from, and some toward the earth) yields the discrepant redshifts.

An even more bizarre complication is hinted at, in a photograph taken by Arp and Kormendy (in *The Redshift Controversy*, edited by Field *et al.*). That controversial plate contains strong evidence of a filamentary linkage between the Stephan cluster and NGC 7331, the apparently much nearer galaxy in our previous illustration. Three smaller galaxies that lie immediately adjacent to NGC 7331 seem also to be enmeshed in the same filamentary network, yet their observed redshifts suggest a velocity in the range of 5000 to 6000 km/sec. In the case of NGC 7331 and its 'satellites' we seem to have an extension of the redshift paradox of Stephan's Quintet.

NGC 7332
'Peculiar' galaxy in Pegasus
RA   22h 35m
Dec +23° 32'
Mag   12.0

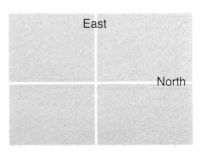

East

North

The visual impression of this system is not unlike that of the other notable Pegasus galaxy, NGC 7331. NGC 7332 is much fainter, however, and requires at least a 20-cm telescope for a minimally satisfactory view.

NGC 7332 has the combination of flattened disc and lack of recognizable spiral structure that characterizes the 'SO' class of galaxies, a type that Hubble regarded as a transitional stage between elliptical and spiral systems. If it could be viewed face-on, it would more closely resemble the rather featureless disc of NGC 404 than an elaborate spiral such as M101.

The Hubble Atlas of Galaxies describes NGC 7332 as 'peculiar' because of an apparent abnormality of its central region. This galactic core is not the usual smooth, convex nuclear bulge; it appears irregular and rather angular in form, with a hint of indentation at the ends of the rotational axis.

A faint background object, the edgewise galaxy NGC 7339, lies adjacent to NGC 7332 in the telescopic field.

NGC 7479
Barred spiral in Pegasus

RA   23h 02m

Dec +12° 03′

Mag   11.5

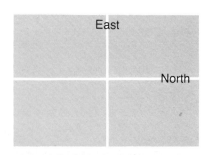

East

North

NGC 7479 is a classic of the S-shaped barred spiral type; it is characterized by a very small condensed core, a pronounced transverse bar and thin, loosely wound spiral arms.

A notable feature of this galaxy is its obvious asymmetry. Whereas the southern half of the central bar graduates smoothly into a single, powerfully curving arm, the northern projection dissolves into a series of weak branches. The most prominent concentration of dust in this oddly lopsided star-system appears among the complexities of the divided northern arm.

Barred spirals such as NGC 7479 have been found to differ from normal spirals in a curious way: the main part of the bar appears to be a region of the galaxy in which stars do not form, although normal condensation of dense protostellar clouds to produce new stars occurs at the outer ends of the bar, and in the spiral arms.

At least a 20-cm telescope is required for a satisfactory glimpse of this rather dim object. With a 40-cm instrument, one may begin to detect the general north–south elongation of the galaxy's transverse bar.

M16 The Eagle Nebula
NGC 6611
Diffuse nebula in Serpens
RA   18h 18m
Dec −13° 48′
Mag approx. 6.5

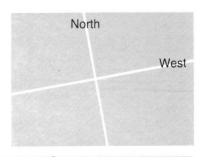

Visual observations of this object with modest optical aid show little or no hint of the nebula itself; the associated star cluster, however, is an attractive handful of bluish jewels. With an aperture of about 30 cm one begins to detect the hazy network of illuminated hydrogen gas in which the stars are enmeshed.

The cluster is a youthful one – perhaps no more than a million years old – and its surrounding nebulosity, like that of the similar stellar nursery M8, contains compact dark globules that may be protostars.

The photograph shows well the intriguing multi-dimensional appearance of M16, with its foreground 'thunderclouds' of obscuring matter looming in sharp relief against the starlit emission nebula behind. Drs Bart and Priscilla Bok (*The Milky Way*, 4th ed, p. 56) describe the tortuous outlines of these dark projections as evidence of shock fronts within a region of great turbulence.

A measurement of distance, based on estimated absolute luminosities of stars within the cluster, is difficult for M16 because of the heavy obscuration of light from the field-stars. If 8000 light-years is accurate, the deduced scale of the nebula is impressive. The principal dark projection is about 6 light-years long (considerably greater than the distance between our Sun and the next closest star), arching part way across the 60-light-year illuminated gulf behind.

M17 The Omega Nebula
NGC 6618
Diffuse nebula in Sagittarius
RA   18h 20m
Dec −16° 12′
Mag   7.0

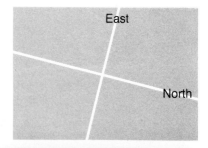

This object is a good one for comparison with M16, for it is a similar region of ionized hydrogen gas excited to luminosity by hot stars within. Visually, it is an easier target than M16; the nebulosity is readily seen with even a very small telescope. Its distinctive shape has suggested various names (the Omega, the Swan, the Horseshoe), all of which seem to be commonly in use.

M17 is a somewhat less distant nebula than M16, but of approximately the same intrinsic size. Although the material of this interstellar cloud is so rarefied that it differs little from a perfect vacuum, its total quantity is substantial – the equivalent of about 30 000 solar masses.

C.J. Lada (*Astrophysical Journal*, vol. 32, pp. 603–29, 1976) has reported recent investigations of a complex molecular cloud associated with M17, showing evidence of an array of chemical combinations that include carbon monoxide and ammonia, among others.

Infrared sources in particular localities within the nebula suggest that this, too, is a region of new star formation.

Although diffuse nebulae, as a class, tend to be disappointing objects when studied visually (most reveal their wispy forms only on long photographic exposures), the Omega is an exception. The amateur astronomer will find the bright, elongated glow of M17 only a little less impressive than the Orion Nebula, or the Lagoon.

M20 The Trifid Nebula
NGC 6514
Diffuse nebula in Sagittarius
RA   18h 01m
Dec −23° 02′
Mag   9.0

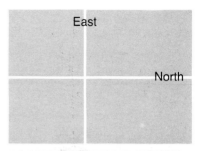

East

North

The Trifid Nebula is so named because of its distinctive three-lobed emission region. On a night of acceptably dark and transparent skies, a 20-cm telescope will show the network of obscuring dust lanes that create the illusion that M20 is divided into three separate masses. Even a very small telescope will reveal HN 40, a bright double star that lies at the central junction of these dark rifts. This giant binary (actually a multiple star involving several fainter companions) appears to be the energy-source by which the reddish emission nebula is excited to luminosity.

The region of M20 that emerges with a blue colouring on the film emulsion is a reflection nebula: unlike the brightly glowing Trifid portion, it is visible to us principally by light reflected from its particles, rather than by light emitted from within.

Although the distance of this object is a subject of controversy, it is possible that the Trifid and the Lagoon Nebula (M8) lie at similar distances, and are parts of a single vast cloud. A low-powered, wide-angle telescope will show both nebulae within the same field.

In 1839, the brilliant young astronomer E.P. Mason undertook the drawing of a series of painstakingly precise isophotal charts of the nebula and its dark rifts, using a homebuilt 30-cm reflector that was at that time the largest telescope in the United States. It was his belief that, if similar drawings were made several decades later, a comparison of them with the 1839 charts might reveal changes in the size and shape of the Trifid.

Although Mason, who died at an age of only 21 years, was unable to complete the project, his superbly executed drawings are still extant (see *Sky and Telescope*, Dec. 1972, pp. 366–7). Comparison with modern photographs shows no detectable change.

M22
NGC 6656
Globular cluster in Sagittarius

RA 18h 35m
Dec −23° 56′
Mag 5.9

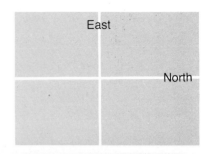

East

North

Of this magnificent object, the 19th-century writer T.W. Webb noted: 'a beautiful bright cluster, very interesting from visibility of components, largest 10 and 11 mag., which makes it valuable for common telescopes . . .'.

One of the nearest of all globulars, the great Sagittarius cluster would dramatically outrank the more famous M13 for amateur astronomers in northern latitudes, if it were not at such a far southerly declination.

A glance in the direction of M22 with the unaided eye will show that it lies in the bright, crowded path of the Milky Way in Sagittarius. The impression of its entanglement among the thronging stars of that region is indicative of its actual situation in space. Unlike many other prominent globulars, which occupy positions high above or below the disc of the Milky Way, M22 hovers very low above the galactic plane.

In spite of its difficult

position for most observers, M22 is an extraordinary sight in a telescope of moderate aperture. Lying at a distance less than half that of the Hercules cluster (M22 is about 10 000 light-years distant) and being rather less condensed than that notable cluster, the Sagittarius cluster is more easily resolved into individual stars. It merits inclusion among the showpiece objects of the night sky in midsummer.

M27 The Dumbbell Nebula
NGC 6853
Planetary nebula in Vulpecula

RA   19h 59m

Dec +22° 40′

Mag   7.6

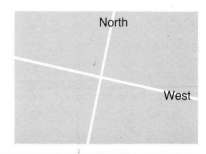

North

West

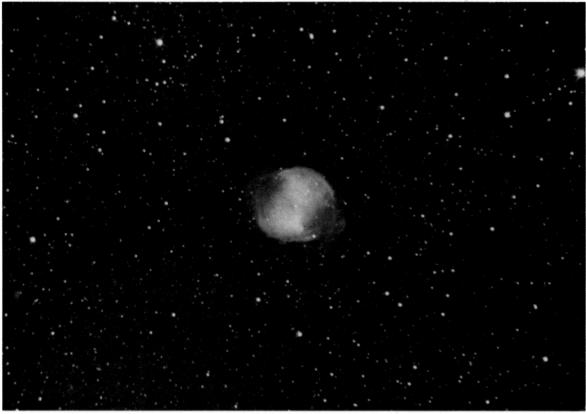

Much brighter, larger and more dramatic than the best known planetary nebula, M57, is the Dumbbell. Its portrait shows the characteristic hour-glass shape of M27, of which some hint is seen with even the smallest amateur telescope.

Spectroscopic and astrometric studies of M27 have provided insights into the Dumbbell's origin and history. A faint star near the centre of the glowing mass, having the characteristic spectrum of a very hot collapsed object, is identified as the source of the vast gaseous shell. Comparative measurement of photographic plates shows a detectable expansion of the nebula; the shell is still flying outward from an explosion that must have occurred several tens of thousands of years ago.

It is useful to draw some comparisons between this typical planetary nebula and the diffuse nebulae that we have seen in photographs on previous pages. In contrast with the 100-light-year width of M8, for instance, M27 is a tiny object; the long axis of the Dumbbell is a little under 3 light-years. While the great emission nebulae lie at relatively remote distances (5000 and 8000 light-years, respectively, for M8 and M16), the Dumbbell might almost be called a close neighbour, at a probable distance* considerably less than 1000 light-years.

*M27, however, is one of those objects whose distance has posed considerable difficulty. While several authoritative estimates have fallen in the range of 600 to 900 light-years, at least one recent source gives a value of 3500 light-years!

M29
NGC 6913
Open cluster in Cygnus
RA   20h 23m
Dec +38° 27′
Mag   7.1

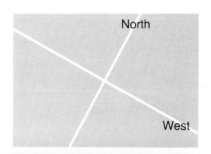

Although very small and sparse, M29 is nevertheless a rather pleasing object for the telescope; to the present writer this cluster seems like a fascinating miniature of the Pleiades (M45).

The comparison between the two star-groups is an interesting one. Dim little M29 is in fact about twice the size of the Pleiades, but more than seventeen times as far removed from us. Its principal members are superluminous bluish-white B-type stars like the famous Alcyone, Merope, Atlas, etc., in the Pleiades.

M29 is an undeservedly neglected open cluster, described as 'poor' in many of the classic observers' handbooks; T.W. Webb's *Celestial Objects for Common Telescopes* omits any mention of it. Good binoculars, however, are all the optical aid required for a glimpse of the cluster, and a 7-cm telescope resolves the neat little 'dipper' shape outlined by the half dozen brightest members. The two tiny stars that lie just within the corners of the cluster's central rectangle are an intriguing test for the visual observer; they are at the limit of visibility in a good 10- or 12-cm instrument.

The Cocoon Nebula
IC 5146
Diffuse nebula in Cygnus
RA   21h 50m

Dec +47° 10'

Mag (faint)

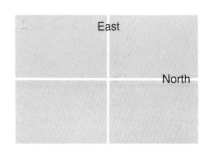

East

North

As an object for the visual observer, the 'Cocoon' is unexciting. A moderately large amateur instrument (25 or 30 cm) shows it as a sparse cluster of 12th-magnitude stars within which there is a ghostly hint of nebulosity. Like many of the diffuse nebulae, it reveals its true aspect only on film exposed for several minutes.

IC 5146 may be regarded as a smaller and more distant version of the great Orion Nebula (M42). It is a region of ionized hydrogen gas excited to luminosity by an energetic, embedded star and, like the famous M42, it is associated with a much larger, but unseen, nebula. At an estimated distance of about 3100 light-years, IC 5146 is about 70 per cent further away from us than the Orion Nebula.

The Cocoon Nebula is aptly named, for within its complex folds of gas and dust, protostellar globules undergo the transformation that gives birth to new stars. The luminous central star itself is a comparatively newly hatched object; it is believed to be considerably less than half a million years old.

The Veil Nebula
NGC 6960, 6992, 6995
Supernova remnant in Cygnus
RA (NGC 6960) 20h 44m

Dec (NGC 6960) +30° 32′

Mags various (all very faint)

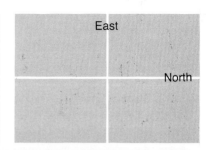

Although these scattered strands of faint nebulosity are too widely separated to be encompassed in a single photograph, they are all parts of a single object in space. Together, the gaseous ribbons of NGC 6960, 6992 and 6995 form one vast, circular shell perhaps as great as 100 light-years in diameter.

Like some more compact shell-type nebulae (cf. M27), the Veil shows a perceptible degree of expansion outward from its centre, when high-precision photographs taken at widely separated intervals are measured. Almost certainly we are observing a gigantic bubble of material ejected by a star that

exploded as a supernova at some moment in the remote past – probably over 100 000 years ago. The more famous supernova remnant M1, the Crab Nebula, is a recent and still relatively small splash of stellar gases; the Veil, by contrast, illustrates the stage at which the bubble, now thinly distended, is

The Veil Nebula
NGC 6992, 6995
Supernova remnant in Cygnus
RA (NGC 6992) 20h 54m
Dec (NHC 6992) +31° 30′
Mags various (all very faint)

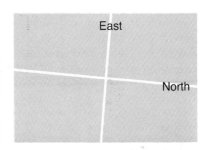

nearing its final dissolution. The ultimate fate of at least some of this material may be formation of new stars. Astrophysicists are fond of reminding us that not only our Sun and its planets, but our bodies themselves are made of elements that once lay inside some earlier-generation star, long dead and gone.

The photographs show the intricacy of colour and detail that characterize the lovely Veil Nebula. The subtle range of hue – wisps of red contrasting with regions of bluish-white – is indicative of temperature differences within the nebula. Visually, however, the object is a difficult test of the observer's skills. On a night of superlatively dark, transparent skies a moderate telescope reveals the delicate curve of NGC 6992 with little difficulty. (The present writer has glimpsed this segment of the Veil with a little 9-cm Celestron telescope.) The much fainter opposite hemisphere of the shell, represented by NGC 6960, requires a larger aperture, in the range above 25 cm.

NGC 6946
Spiral galaxy in Cepheus
RA   20h 34m

Dec +59° 58′

Mag   11.0

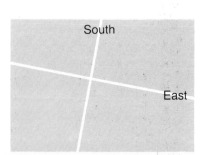

NGC 6946 is a very loosely wound spiral galaxy, viewed almost directly face-on. Although visually not a prominent object (its general surface brightness is extremely low), it seems to be a relatively nearby neighbour to our own galaxy; at a probable distance of 10 or 15 million light-years, it lies not far beyond the compass of our Local Group.

In the long, complex spiral arms, major knots of luminous hydrogen gas are resolved and, on several occasions in the past century, supernovae have been identified.

NGC 6946 is a star-system characterized by a small, bright, compressed nucleus and rather diffuse arms. A 20-cm telescope gives an elusive glimpse of only the galaxy's central region. Reports by visual observers with larger instruments (*Webb Society Deep-Sky Observer's Handbook*) indicate that apertures up to 40 cm reveal little more detail in this dim, although very large, object.

M30
NGC 7099

RA   21h 39m
Dec −23° 15′
Mag   8.4

At magnitude 8.4, M30 is a considerably less prominent globular cluster than the spectacular M13 or M22. It can be located with good binoculars adjacent to the star 41 Capricorni. It is about 40 000 light-years distant.

NGC 6804

RA   19h 29m
Dec +09° 07′
Mag   12.0

Comparable with the famous Ring Nebula (M57), this dim planetary nebula is the small, spherical shell of material ejected from a collapsing central star. While M30 is a huge object – perhaps 100 light-years in diameter – the planetary is only a light-year or so in breadth.

M31 The Andromeda Galaxy

NGC 224

Spiral galaxy in Andromeda

RA   00h 42m

Des +41° 09′

Mag   4.8

All of the celestial objects portrayed in the present book have been photographed on the same scale. If one bears this fact in mind while comparing the image (in two parts) of M31 with those of many other galaxies, an impression of the Andromeda system's great size and relative nearness to us in space emerges.

Visible as a dim oval glow to the naked eye, M31 was an enigma to astronomers for centuries. In the 1920s Edwin Hubble finally demonstrated that we are looking, not at a small object lying within our own galaxy, but at an enormous system comparable to our own Milky Way, situated far beyond our galaxy's boundaries. The key to Hubble's solution of the longstanding question of the Andromeda object's distance was the discovery in 1923 of individual stars within M31 of the Cepheid variable type.*

Modern measurements show that the Andromeda Galaxy is an enormous system of more than 300 billion stars, and that, with a diameter of over 150 000 light-years, it is a considerably larger cousin to our own spiral galaxy. Its estimated distance (2.2 million light-years) seems prodigious. On the galactic scale, however, on which similar star-systems are seen to be hundreds or even thousands of millions of light-years away, M31 is a mere backyard neighbour.

M31 is a spiral galaxy whose appearance, if we could observe it directly face on, would be roughly comparable to that of M58 or M101. As it happens, the Andromeda Galaxy is viewed from the earth at an oblique angle, with the plane of its disc inclined only about 12 degrees from the edge-on orientation.

The two photographs reveal many of the components that are typical of a spiral galaxy similar to the Milky Way. In spite of difficulties presented by the very oblique angle of M31, the structure is clearly seen, including such easily recognized

40

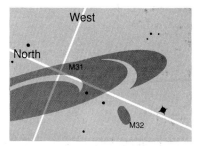

details as the dense compact nucleus, the spiral arms and lanes of dark obscuring matter, and small, bright knots of luminous material – the so-called H II regions of hot stars embedded in clouds of ionized hydrogen gas.

Our Milky Way system has smaller galaxies gravitationally associated with it in space (we recognize the Greater and Lesser Magellanic Clouds as satellites of this kind). Similarly, the two small 'nebulosities' that can be seen close to M31 in the photographs, and even visually with moderate telescopes, are actually miniature galaxies connected with the Andromeda

system. M32 and NGC 205 † are small elliptical galaxies situated close enough to the massive primary galaxy to be tidally distorted by its gravitational influence. Paul W. Hodge, in his excellent *Galaxies and Cosmology*, describes the complex interaction of M31 and M32; details of the configuration of the Andromeda Galaxy's spiral arms appear to have resulted from the deforming influence of little M32 nearby.

The Andromeda Galaxy is by far the most easily observed object of its kind. Binoculars show both the bright central core and the hazy disc of this huge star-system. A 7-cm telescope at low magnification will pick up the

faint images of the small elliptical companions. It is a peculiarity of M31, however, that it yields little further detail when observed visually with larger telescopes.

* See the page of text on M3 for elaboration of the astronomer's use of these variable stars as indicators of distance in space.

† NGC 205 is sometimes added to Charles Messier's list, and given the number M110.

41

NGC 404
'Peculiar' galaxy in Andromeda
RA   01h 07m

Dec +35° 27'

Mag   11.5

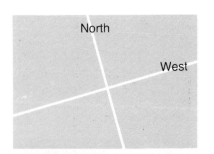

North

West

Intrinsically dim, and located immediately adjacent to the glaring brilliance of 2nd-magnitude β Andromedae, this object is an inordinately difficult challenge to the visual observer.

When Edwin Hubble developed the first scheme of classification for observed galaxy-types, he identified a class that seemed to be the transitional stage between elliptical and spiral types. Galaxies of this 'S0' class are characterized by a generally featureless ovoidal form, without spiral structure, like the ellipticals; they exhibit also a sparse surrounding envelope (often flattened into a disc) vaguely reminiscent of the outer parts of the spirals' anatomy.

Although NGC 404 is normally catalogued as one of these 'S0' systems, it has also been described as 'peculiar' because of a pronounced, arc-shaped dust lane that outlines one half of its quite prominent nucleus.

The portrait shows this galaxy's circular shape and apparently face-on orientation. It is believed, however, that these impressions are misleading; NGC 404 is more likely a rather elongated form, tilted at some angle away from our line of sight.

NGC 185
Elliptical galaxy in Cassiopeia
RA   00h 38m

Dec +48° 04′

Mag   11.7

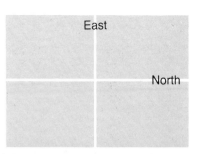

East

North

Extremely difficult to locate and observe with an amateur telescope, this pale little object is a small elliptical galaxy whose true linear size is less than a tenth that of our own galaxy.

NGC 185 has a number of associated globular clusters. The fact that these clusters' angular diameters compare closely with diameters of globulars in M31 (the Andromeda Galaxy) was an early indication that the small elliptical lies at a distance roughly equivalent to that of the great spiral.

Although it is a full 7 degrees away from M31 in the sky – a separation of over 250 000 light-years in actual space – NGC 185 is a satellite of the larger galaxy. In fact the Andromeda system together with its family of attendant dwarf elliptical galaxies (including M32, NGC 147 and NGC 205) almost constitutes a cluster in its own right; these five have been described as a 'subunit' of the Local Group of galaxies.

Being a nearby neighbour to us in intergalactic space, NGC 185 is resolved into its myriads of component stars, in photographs made with the world's largest telescopes. To its great pioneer observer Walter Baade, it had the appearance of a gigantic, supermassive globular cluster.

A feature that makes this dwarf galaxy atypical among ellipticals is the significant amount of dust among its stars, especially in the form of unusual obscuring globules that have been observed near the core of the system.

NGC 278

RA   00h 49m
Dec +47° 18′
Mag 10.5

This galaxy is located just east of NGC 185. Appearing circular and almost featureless, it is in fact an exceedingly compact, tightly wound spiral, viewed directly face-on. At magnitude 10.5, it is a difficult target for any instrument under about 20 cm.

NGC 1245

RA   03h 11m
Dec +47° 03′
Mag 9.0

A 9th-magnitude open cluster of about 100 members, this misty patch is easily located 3 degrees southwest of α Persei.

NGC 891
Spiral galaxy in Andromeda
RA   02h 19m

Dec +42° 07′

Mag   10.5

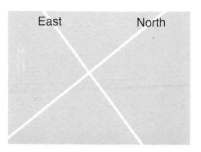

About 40 million light-years distant, this very large spiral galaxy would be a considerably brighter and more impressive sight in amateur telescopes if it were viewed face-on. Oriented directly edgewise to our line of sight, it is heavily masked by its unusually broad equatorial dust lane.

In many galaxies the dust is confined, by gravitational effects, to a relatively thin disc on the plane of rotation. The central lane of NGC 891, by contrast, is greatly distended north and south of the galactic equator, with numerous peaks and prominences that spread far beyond the central rotational plane. Astrophysicists regard this as evidence of high levels of energy being 'pumped' into this galaxy's equatorial dust-region.

NGC 891 is a relatively complex, loosely wound spiral, regarded as probably similar in most aspects to our Milky Way. It is interesting to compare it also with NGC 4565, a dramatic and well known telescopic object which it closely resembles.

NGC 7662
Planetary nebula in Andromeda
RA   23h 23m

Dec   +42° 12′

Mag   8.5

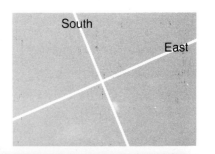

This elegant little gem is a compact planetary nebula, one of the easiest to observe with a telescope. Although much smaller than the famous Ring Nebula in Lyra (M57), the little-known NGC 7662 is brighter. A good 8-cm refractor at a magnification of about 75 × will resolve this object as a disc, almost circular, with a pale bluish-green tint.

The visual appearance of NGC 7662 with a larger instrument is described in the 19th-century notebooks of T.W. Webb: 'My 9¼-inch shows bluish disc with woolly border and suspicion of dark centre.' Modern observations confirm the impression of an incipient ring formation; like many planetaries, this nebula is actually a bubble with a relatively vacuous centre. A blue dwarf star within, probably the source of the ejected shell of gas, is an unconfirmed variable suspected

of occasional bursts that increase its brightness by several magnitudes.

The impression that this is a small planetary nebula is accurate. The probable diameter of its outer circle is about half a light-year – perhaps one-fifth the size of the impressive Dumbbell Nebula (M27). The limited expansion of the gas into space indicates a relatively recent date for the event that created NGC 7662.

M33
NGC 598
Sc galaxy in Triangulum
RA   01h 33m
Dec +30° 33′
Mag   5.8

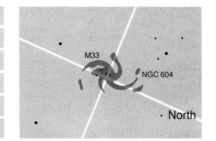

The dim, but extensive, face of this excellent spiral – visually, larger than the full moon – gives the impression that M33 is a relatively near neighbour in intergalactic space. The impression is accurate: lying only a little more than 2 million light-years distant, the Triangulum galaxy is part of the Local Group, approximately as far away from our own galaxy as M31 in Andromeda.

Of the seventeen closely associated galaxies that constitute the 'Local Group', only the three mentioned in the preceding paragraph are full-sized spiral galaxies. The triangle thus formed in space is rather interestingly shaped, in that the separation of M33 from M31 is relatively small – only a quarter of the distance between our own galaxy and either of these other two.

Although M33 is smaller than our own Milky Way system (less than three-quarters our galaxy's diameter and perhaps one-twentieth its mass), it is the easiest of all the spiral galaxies for study with our telescopes. This is because of its fortuitous combination of proximity, nearly face-on position, and looseness of structure. Great instruments such as the 5-m Hale telescope at Palomar resolve this spiral's principal clots of nebulosity, the open star clusters strung along its S-shaped arms, and teeming individual stars.

Several large, bright nebulae within M33 are sufficiently distinctive to have been assigned NGC numbers in their own right. A 20- or 25-cm telescope will show the most notable example, a local condensation similar to the well known Orion Nebula in our own galaxy: NGC 604 can be identified in the northern arm about 10 arc-minutes from the Triangulum galaxy's centre.

M34

NGC 1039

RA   02h 41m

Dec +42° 43′

Mag   5.5

An easy binocular object, M34 is a stellar family of approximately 80 members, within a space 4 light-years in diameter. Like M36, this is a cluster of very young stars.

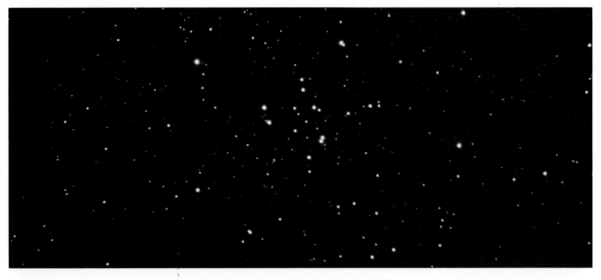

NGC 1023

RA   02h 37m

Dec +38° 52′

Mag   10.0

Located about 3 degrees directly south of M34, NGC 1023 is a highly elongated elliptical galaxy whose bright nucleus and tenuous fringes can be differentiated with a 15-cm telescope.

M35
NGC 2158
Open cluster in Gemini
RA   06h 08m

Dec +24° 21′

Mag   5.3

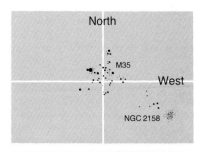

The photograph shows M35 and a much fainter companion-object NGC 2158, which is in reality an extremely distant background cluster.

The principal cluster is one of the finest in the sky; with an apparent size equivalent to the full moon, it is a spectacular sight in binoculars or a small telescope. The 30-light-year diameter of M35 encompasses a swarm of at least 300 stars, the brightest of which are hundreds of times as luminous as the Sun. The characteristic yellow and orange colours of these G- and K-type giants are recorded in the photograph, and can be easily detected when the cluster is observed visually through the telescope.

It is interesting to note that, when we study this compact group, we are looking at a small region of space in which the typical distance from one star to another is much less than in our Sun's stellar neighbourhood. In our own vicinity, the average density of matter is about one-tenth of a solar mass per cubic parsec; M35 crowds perhaps ten times that amount of material into the same volume of space.

A comparison of the two clusters in the photograph also provides intriguing insights. We are gazing into an enormous depth of field; while M35 is in the relative foreground at a distance of 2200 light-years, NGC 2158 lies in the far background – about 16 000 light-years distant! This vastly remote object, situated out near the galactic rim, is a puzzling cluster that perhaps represents a transitional phase between the open and the globular types.

49

The Eskimo Nebula
NGC 2392
Planetary nebula in Gemini
RA	07h 28m
Dec	+20° 57′
Mag	8.3

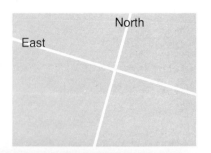

The photograph shows the bright central 'face' and much fainter surrounding fringe that suggest the nickname 'Eskimo Nebula' for this little planetary. Exposures made through the world's largest telescopes (particularly in red light) bring out traces of dark features on the central disc that look rather eerily like eyes, nose and mouth.

The probable distance of this compact bubble of gas is of the order of 3000 light-years. Its small intrinsic size – comparable to the Ring Nebula in Lyra – suggests that this envelope of material was expelled from its central star quite recently; the age of the nebula is likely to be under 2000 years.

A recent photometric survey of NGC 2392 (R. Louise, *Astrophysics and Space Science*, vol. 79, pp. 229–37, 1981) indicates an absence of the classical stratification structure observed in planetary nebulae. Instead, the author of the report suggests an unusual model for this object, consisting of an inner toroid surrounded by a spherical outer shell.

The Eskimo Nebula is unjustly neglected by many amateur observers. Although very tiny, it is a bright and easily located target for even the smallest telescopes. Through a 15-cm aperture the nebula is a featureless, but distinctly turquoise-coloured, disc of light. A hint of the faint outer ring may be glimpsed with a 20- or 25-cm instrument.

M36
NGC 1960
Open cluster in Auriga
RA   05h 35m
Dec +34° 05′
Mag   6.3

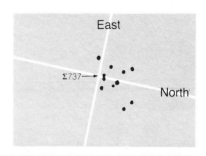

M36 is one of three superb star clusters that can be swept up with binoculars, all within one 5-degree field in the autumn constellation Auriga. It is a relatively sparse, loosely knit stellar association whose few dozen members are spread over a small locality in space measuring about 12 light-years in diameter. The distance of the group is of the order of 4000 light-years.

Although small and distant, this cluster is very bright – a sparkling knot of bluish diamonds, when viewed with a 5- to 8-cm refractor. Its prominence is due to the high intrinsic luminosity of the very hot, brilliant B-type stars that constitute much of its population.

A small telescope reveals that the star labelled 'Σ737' us a rather easily resolved double.

It is instructive to compare this loose and very open cluster with the strikingly different neighbouring object M37, illustrated on the following page.

M37
NGC 2099
Open cluster in Auriga

RA   05h 52m

Dec +32° 33′

Mag   6.2

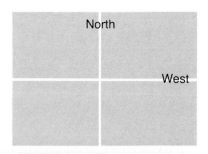

In contrast with M36 (preceding page), this is an extremely populous and densely compressed cluster. Within its 20-light-year circle it crowds several hundreds of stars that can be resolved in photographs like the one above, taken with the 40-cm reflector; undoubtedly there are hundreds of smaller members too faint to be detected at the cluster's 4200 light-year distance.

In a 7-cm telescope M37 is a most intriguing object, having the appearance of a nearby and easily observed globular cluster.

Of all the open clusters, only M11 in Scutum is a rival to this throng of tightly packed stars. Ranging in visual magnitude from 9th down to about 13th magnitude, the myriads of individual cluster members are resolved by a small telescope as a dense scattering of fine dust.

In *Celestial Objects for Common Telescopes*, T.W. Webb comments on an early cosmological insight that was provided by M37. At a period in astronomical history when it was supposed that all stars were of approximately the same

absolute luminosity, most observers accepted the corollary that all stars that appear relatively dim are relatively distant. Bright stars were thought to be closer to us. M37, long recognized as a true family of stars all lying at the same distance from the earth, exhibited an obvious range of brightnesses among its members. Thus, the cluster was a demonstration of the fact that apparent luminosity is not a function of distance alone; some stars are intrinsically much less luminous than others.

M38
NGC 1912
Open cluster in Auriga
RA   05h 27m
Dec +35° 48′
Mag   7.4

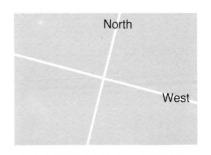

M38 completes the trio of galactic clusters in Auriga. Like the other two groups, it lies in a region somewhat over 4000 light-years distant from us.

Although this is a relatively dim and scattered cluster, it contains a number of quite conspicuous stars –B- and G-class giants having absolute luminosities hundreds of times that of our Sun. In contrast with even an undistinguished cluster like M38, the neighbourhood occupied by the Sun is indeed a poorly lit sector of space. The sparse association of about sixty stars that comprise the 'Local Group' are mostly specimens of a type very different from the energetic giants of the Auriga clusters. We live in the midst of a feeble throng whose members are chiefly red dwarfs, a few thousandths as luminous as the Sun itself.

M38 is worth observing, particularly for the challenge offered to small apertures by the much fainter and smaller cluster that lies in the same field. NGC 1907, which can be seen adjacent to the southern fringes of M38, is a similar cluster situated in the distant background.

NGC 1931

RA   05h 28m
Dec +34° 13′

A hazy, nebulous patch of illuminated hydrogen gas, just 3 degrees south of M38, this object can be detected visually with a 20-cm telescope.

NGC 7217

RA   22h 56m
Dec +31° 07′
Mag 11.0

At 11th magnitude, the galaxy NGC 7217 is not an easy target for small telescopes. It is a rather odd spiral with one long, tightly wound arm that totally encircles the galactic core.

IC 405
Diffuse nebula in Auriga
RA   05h 13m
Dec +34° 16′
Mag (very faint)

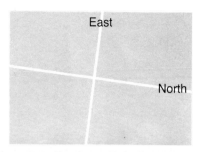

It is worthwhile to compare this photograph with that of the famous Orion Nebula (M42); both of these interstellar clouds lie at very roughly comparable distances from us: about 1800 light-years for the Orion Nebula and 1600 in the case of the Auriga object. As their portraits suggest, IC 405 is a considerably smaller region of nebulosity than M42 – approximately 9 light-years in overall breadth.

IC 405 displays a strikingly filamentary structure, with bright lobes of glowing hydrogen gas bordered by clouds of obscuring dust.

The bright star on the nebula's edge is not a foreground object, but is actually embedded in the dusty, gaseous fold of IC 405, and appears to be the nebula's source of illumination. Designated AE Aurigae,* this star is an erratic variable, subject to intermittent surges and diminutions of brightness; the star ranges between 5th and 6th magnitude.

IC 405 is one of those regions of nebulosity that elude visual detection with small telescopes, but which reveal their presence on film exposed for several minutes.

* Blaauw and Morgan (*Astrophysical Journal*, vol 119, 1954) have observed that the star AE Aurigae appears to be in transit through the nebula, having encountered this interstellar cloud after being ejected at high velocity from the Orion Association.

M39
NGC 7092
Open cluster in Cygnus
RA   21h 32m
Dec +48° 21'
Mag   5.2

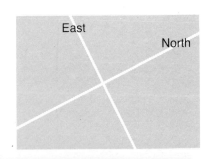

This course, scattered group makes an interesting contrast with the clusters M36 and M38 on previous pages. Similar to them in size and stellar content, it appears many times larger and brighter because of its nearness to us in space. Its distance is under 1000 light-years, or less than a quarter that of M36 or M38.

Approximately thirty stars are members of this sparse stellar family, most of them hot, superluminous A- and B-type suns whose characteristic blue-white glare is noticed in the photograph, and in visual observation of the cluster.

Its very large apparent size (nearly twice that of the full moon) makes M39 a difficult object to appreciate with ordinary astronomical telescopes. Binoculars, on the other hand, show these lovely stars as a compact cluster; a 15-cm rich-field telescope at very low magnification yields a stunning view, with easy resolution of at least twenty cluster-members. M39's brightness makes it a useful early-evening target for star parties and public viewings on summer evenings when the sky remains too light for advantageous display of more subtle deep-sky objects.

M42 The Orion Nebula
NGC 1976
Diffuse nebula in Orion
RA   05h 34m
Dec −05° 24′
Mag approx. 4

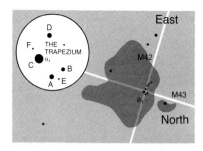

With the unaided eye most observers easily notice the mistiness of the middle star of Orion's sword. Binoculars enhance this impression of nebulosity, and a good telescope reveals the complex splendour of the Orion Nebula, with its quadruple central star, invading black dust lanes and branching luminous filaments. Immediately adjacent to M42, the smaller M43 (NGC 1982) is a detached part of the main nebula that will also be seen visually with the small telescope.

M42 is an enormous cloud of gas and dust, about 30 light-years in actual diameter and probably of the order of 1800 light-years distant from us. This remarkable nebula is regarded by astrophysicists as a sort of celestial nursery in which the process of stellar birth is taking place even as we watch.

The central star, the 'Trapezium' (designated θ_1 Orionis) is, in fact, a little cluster of very young stars that have coalesced and become luminous in the relatively recent past. These stars are believed to be less than half a million years old; thus it can be said of this unique cluster that the ancestors of modern man already walked the earth before the Trapezium sprang into life within the Orion Nebula.

The Trapezium is a popular target for amateur astronomers' telescopes, not only because is it one of the most beautiful multiple stars in the heavens, but also because of two additionally interesting features. One of these is the variability of the faintest of the four companion-stars ('B') – an eclipsing binary star whose luminosity decreases noticeably for a few hours at $6\frac{1}{2}$-day intervals. The other intriguing feature is the presence of at least two additional, faint cluster members ('E' and 'F'), the detection of which has long been regarded as a test of both telescope and observer.

NGC 1973–1977

Diffuse nebula in Orion

RA 05h 33m

Dec −04° 48′

Mag (faint)

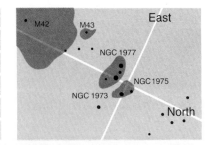

This complex interweaving of bright and dark material surrounds the star 42 Orionis, half a degree north of the Trapezium in the Great Orion Nebula. There is sometimes an element of confusion about the identity of the object; photographs of NGC 1973–1977 occasionally appear in books and journals, labelled 'M43'. (The latter is a considerably smaller nebula attached directly to the Great Nebula's northern fringe, and surrounding an 8th-magnitude star only 7 arc-minutes removed from the Trapezium.)

Like the several other nebulae that crowd this part of Orion, NGC 1973–1977 is an outcropping of the same vast gaseous cloud whose most spectacular lobe is M42. The photograph shows clearly some parts of this billowing mass of dust and gas that display the reddish hue of material illuminated from within; other segments have the typically bluish colour of reflected starlight. On the nebula's southern edge, particularly, the luminous hydrogen gas is masked behind a foreground dust-cloud.

Although neglected by amateur observers, because of the distractions of glorious M42 nearby, NGC 1973–1977 is worth locating and viewing with a telescope. With a 20- to 25-cm instrument, the nebulosity is an unmistakable presence, having a pale greenish hue when observed visually. Some amateurs have detected the brighter parts with apertures as small as 10 cm.

The Horsehead

B 33

Obscuring nebula in Orion

RA   05h 39m

Dec −02° 32′

Mag (faint)

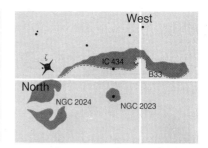

This photograph includes the most complex aggregation of nebulae that can be observed in a one-degree telescopic field. The Horsehead, a spectacular dark particle-cloud, lies silhouetted against the bright nebula IC 434, which is a background object. Immediately east of the brilliant white star ζ Orionis we see another partially obscured nebula NGC 2024. Nearer to the Horsehead itself, NGC 2023 is a circular gaseous region illuminated by its central star.

When it was first detected on a photographic plate in 1889, the Horsehead was thought to be a gap in the bright cloud of gas. Nowadays, however, it is recognized as a billowing mass of dust, lying perhaps some hundreds of light-years nearer to us than the lighted backdrop of IC 434. At a distance of about 1200 light-years, the great head has a true size of approximately a light-year from mane to nose!

Although its photographic appearance creates an impression that the Horsehead is a very dense object, it is in fact a region more devoid of matter than the best vacuum that can be achieved in a laboratory.

How can such empty space obscure background-light so totally? J.M. Greenberg ('Interstellar grains', *Scientific American*, Oct. 1967) has shown that, perhaps rather surprisingly, this effect is what we should expect of exceedingly tiny grains of matter; the greatest obscuration is produced by particles a little smaller than the wavelengths of the incident light. If the Horsehead were a cloud of objects the size of basketballs, its light-obstructing efficiency would be less by a factor of 300 000.

With even a relatively large amateur telescope, superb sky-conditions and a highly trained eye are needed for even a glimpse of the eerie and elusive celestial horse.

The Rosette nebula
NGC 2244 and 2237
Open cluster and diffuse nebula in Monoceros
RA 06h 30m
Dec +04° 54′
Mag (NGC 2244) 6.2

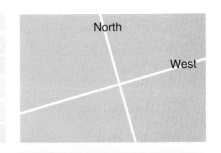

The Rosette presents one of the best demonstrations of the superiority of the camera over the eye in detecting some kinds of deep-sky objects. Visually, the telescope usually reveals only the central cluster NGC 2244 – a bright knot of frosty blue-white diamonds. The truly glorious feature of this group, its surrounding halo of billowing reddish nebulosity, materializes on film during a prolonged exposure.

Like other very massive gas-clouds encountered earlier in this book (cf. M8, M16), NGC 2237 is dotted with the compact dark globules of material that many astrophysicists regard as protostars, collapsing toward the density at which eventually they will begin to emit heat and light. The hot, young O- and B-class stars of the central cluster NGC 2244 are themselves recently created products of the nebula.

The Rosette has an apparent size of nearly one degree (about twice the diameter of the full moon); its true extent is over 50 light-years, 70 per cent larger than the more famous Orion Nebula, which is situated nearby in the winter sky. The Rosette's distance is on the order of 3000 light-years.

The intricate structure of NGC 2237, with its rose-petal form and heavy dark veins of obscuring material, gives the impression of matter condensed to considerable density. This is, of course, an illusion; it is useful to bear in mind that the nebula, although enormously massive, is so tenuous and diffuse as to be little different from a 'perfect' vacuum.

Hubble's Variable Nebula
NGC 2261
Diffuse nebula in Monoceros

RA   06h 36m

Dec +08° 46′

Mag   10 (variable)

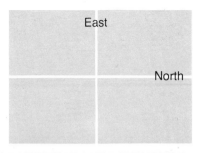

East

North

The visual appearance of this little nebula – a ghostly little wisp of light with slightly brighter apex – is almost exactly like that of a faint telescopic comet. In fact, stumbling upon this object one night while sweeping the region with his 40-cm reflector, Jack Newton felt the thrill of discovering a comet. Having sped home to make a telephone report of his find, he was disgruntled to see, at the new 'comet's' co-ordinates on his Antonin Becvar sky-atlas, the nebula NGC 2261.

NGC 2261 is one of the most peculiar nebulae in the heavens. Early in the present century, Edwin Hubble found that the nebula was significantly changed in separate photographs taken over intervals as small as a few months. Not only did the embedded variable star (R Monocerotis) alter in brightness, but details of the gaseous cloud alternately appeared and vanished. The dramatic effects observed in Hubble's Nebula seem to result from a combination of the erratic illumination provided by R Monocerotis, and changes of lighting-angles caused by actual movement of gas- and dust-filaments lying immediately adjacent to that star.

NGC 2261, lying only about half a degree southwest of the Cone Nebula (see NGC 2264), is believed to be a part of the same vast, general region of nebulosity. At the distance of that complex of gas and dust – 3000 light-years – a rapid series of events observable by us on the earth is a dramatic reminder that remote objects in space are not fixed or static.

Includes the Cone Nebula

NGC 2264

Open cluster and obscuring neb., Monoceros

RA   06h 38m

Dec +09° 56′

Mag   6.0 (cluster)

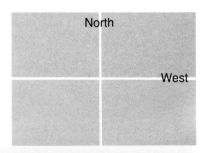

North

West

For an aesthetic appreciation of the unusual cluster NGC 2264 one must use a rich-field telescope with a field of at least 2 full degrees. At a magnification of only 30 ×, the present writer's fine old 12-cm refractor shows a sprawling triangle of forty dazzling stars. Robert Burnham Jr, notes that the cluster has been nicknamed 'the Christmas Tree',

The higher magnification and narrower field of a large instrument loses the overall context of NGC 2264, but permits photography of the region's most intriguing component, the Cone Nebula. This 6-light-year tower of opaque, obscuring dust is oddly similar to the central 'turret' in the nebula M16. Like that region, the huge interstellar cloud that includes the Cone is a stellar nursery, where condensation of nebulous material continues to give rise to newly luminous young stars. Members of the cluster itself are stars that began to emit light as recently as one million years ago. A spectroscopic survey of this region (G.W. Marcy, *Astronomical Journal*, vol. 85, pp. 230–4, 1980) revealed a large number of new T-Tauri variables, a phenomenon often associated with star-generating nebulae.

The location of the Cone Nebula is just at the southward pointing apex of the cluster NGC 2264. The nebula is typical of its dark, silhouetted class in that it is virtually impossible to detect visually.

M44
NGC 2632

RA   08h 39m
Dec +20° 04′
Mag   3.7

The 'Beehive Cluster', one of the most nearby of such groups, at 590 light-years, is visible to the naked eye and easily resolved with binoculars.

NGC 3185, etc.

RA   10h 15m
Dec +22° 00′
Mags All faint

Located due east of M44, in northern Leo, this compact group includes the barred spiral NGC 3185 (southernmost), edgewise spirals NGC 3190 and 3187, and the elliptical galaxy NGC 3193 (northernmost).

M45 The Pleiades

Open cluster in Taurus
RA   03h 46m
Dec +24° 03′
Mag   1.4

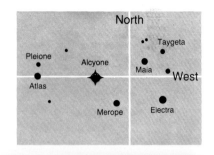

The Pleiades, or Seven Sisters, are the best known star cluster, a favourite object in the autumn sky for naked-eye observers since ancient times. This bright, coarse grouping of stars is a true physical association, one of the most nearby of the galactic clusters, at a distance of only about 410 light-years.

A part of the Pleiades' fascination arises from the fact that, unlike many other notable open clusters (cf. M35, M44), the group appears to the unaided eye not as a vague nebulous patch of light, but as a brilliant nexus of individual stars.

The Pleiades are very young stars, some of which are believed to be still in the process of condensing within the cluster's overall envelope of dust and gas.* The bluish globs of haze seen in the photograph are patches of this overall nebulosity, shining with the reflected light of the embedded hot B-type giant stars (identified by name in the chart).

Bart and Priscilla Bok have written a thought provoking speculation about this lovely young cluster. Because it is a looser and sparser stellar family than some of the very dense groups (e.g. M11 or M37, with their tightly packed thousands of members), M45 is more seriously subject to disruption as a result of encounters with other stars of the galaxy. Calculating the effects of such disruptive encounters, the Boks predict a relatively brief life-expectancy for the Pleiades as we know them. Dismemberment of the cluster, through escape of outlying members and collapse of the core-stars, will significantly alter the Pleiades after a few galactic revolutions; the infant cluster will be entering senility within only five cosmic years.

---

* Recent studies present evidence that contradicts this traditional view of the Pleiades. It is now suspected that this is an older cluster than formerly believed, and that the surrounding nebulosity is not a source of new stellar condensation (*Astronomical Journal*, vol. 87, p.1507, 1982).

M46
NGC 2437
Open cluster in Puppis
RA   07h 41m
Dec − 14° 46′
Mag   6.0

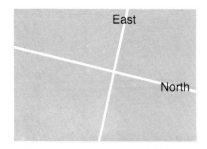

Not only is M46 an outstandingly lovely cluster, but for the amateur astronomer it is made doubly intriguing by the fact that it embodies an object within an object. The portrait of this massive and compressed star-family shows clearly the wispy, nebulous spot (NGC 2438) that surrounds one of its northern members.

Spectroscopic surveys of probable members of this cluster have yielded evidence that the planetary nebula surrounds a star that is not actually part of M46, but a foreground object, much nearer to us in space. The cluster itself, a very dense one having many hundreds of stars, is over 5000 light-years distant.

It is worthwhile comparing this open cluster with a typical globular such as M13 in Hercules. Although the space within the thronging central region of M46 appears to be virtually filled with stars (and has, in fact, almost ninety times as many solar masses per cubic parsec as a similar space in the neighbourhood of our Sun), it is a relatively sparse star-swarm compared with the great globular. In the core of the Hercules cluster, a given unit of space contains a crush of matter nearly four times as dense as that in M46.

Visually, this cluster is a treat, especially at low magnification. At only about $20 \times$ or $30 \times$, it is a glowing circular mist of stars occupying as much of the telescopic field as a full moon. At higher power, an aperture as small as 15 cm permits a glimpse of tiny NGC 2438, against a dazzling backdrop of cluster-stars.

M47
NGC 2422

RA   07h 36m
Dec − 14° 27′
Mag   4.5

Only 1½ degrees west of M46, this is a brighter and looser cluster, suitable for observation with binoculars. Of the two objects, M47 is much the closer, at a distance of 1500 light-years.

NGC 2245 and 2247

RA (NGC 2245) 06h 33m (NGC 2247) 06h 34m
Dec (NGC 2245) + 10° 10′ (NGC 2247) + 10° 20′
Mag (NGC 2245) 8.5

These two small, luminous spots of diffuse nebulosity in Monoceros are bright lobes of the same great hydrogen cloud that includes both the Cone Nebula and Hubble's Variable Nebula. NGC 2245 (southernmost) can be detected with a 20-cm telescope; NGC 2247 is considerably more elusive.

M49

NGC 4472

Elliptical galaxy in Virgo

RA   12h 29m

Dec +08° 07′

Mag   8.6

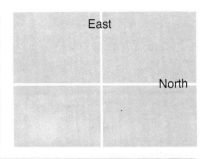

East

North

At visual magnitude 8.6, this is one of the brightest members of the Virgo cluster of galaxies – a comparatively easy object for the telescope.

M49 is linked with a number of other well-known galaxies (M58, M59, M60, M84, M87, etc.) in a vast and relatively nearby galaxy-cloud. With a distance from us of approximately 40 million light-years, M49 lies at about the mean distance of the cluster as a whole.

Although to Charles Messier, observing with a small refractor in 1781, the galaxy appeared to be no more than a nebulous spot, it is in reality one of the most gigantic of the elliptical galaxies. Its mass (equivalent to 1000 billion suns) is five times that of our own Milky Way galaxy.

A newcomer to astronomy who wishes to sample the galaxies of the Virgo Cluster will find M49 a suitable first target on which to test his finding skills. In a 15-cm telescope it is seen as a bright, dense spot surrounded by a periphery of fainter light.

NGC 4535
Spiral galaxy in Virgo
RA 12h 32m
Dec +08° 28′
Mag 10.7

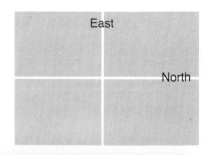

East

North

Located just one degree east of the brighter and much better known M49, this rather pale spiral galaxy is actually associated with the giant elliptical in space, as a member of the Virgo Cluster.

Oriented at an angle that is not far from directly face-on, NGC 4535 displays a symmetrical S-shaped pattern very similar to that of the classic barred spiral

NGC 7479 in Pegasus. The Virgo spiral is, in fact, classed as a barred galaxy in most catalogues. It is structurally quite distinctive, with a small bright nucleus and sparse, spidery arms.

Visually, NGC 4535 is an elusive object. Because of its low overall surface brightness, it has been called the 'Lost Galaxy' in observing lists by Copeland and

Burnham. The amateur astronomer, sweeping eastward from M49 with a rich-field telescope, may easily confuse this dim spiral with a close neighbour; lying only half a degree to the south, NGC 4526 is a moderately bright elliptical galaxy. Half a degree to the north is yet another object, the faint (12th-magnitude) spiral galaxy NGC 4519.

M51 The Whirlpool Galaxy
NGC 5194
Sc galaxy in Canes Venatici
RA   13h 29m

Dec +47° 18′

Mag   8.1

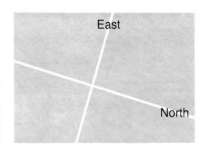

This noble face-on spiral and its (seemingly) connected companion-galaxy comprise one of the best known deep-sky objects, 'The Whirlpool'.

The main galaxy of the pair is the first in which the existence of a spiral structure was recognized. A widely published drawing made in 1850 by the Irish astronomer Lord Rosse showed many of the features seen in the photograph, including the two separate galactic nuclei, the convoluted arms and dust lanes, and the H II regions – small, bright knots of ionized gas.

Situated at a distance of 37 million light-years, M51 is a galaxy of only moderate size; its mass of approximately 160 billion suns is considerably smaller than that of our nearer neighbour M31 in Andromeda. Surprisingly, the companion-galaxy (NGC 5195), which is clearly much smaller than M51, is believed to be about twice as massive – probably a dense and compact elliptical galaxy.

The visual appearance of M51 raises an immediate question: is NGC 5195 actually attached to the larger galaxy's northeast arm? In fact, an elongation of the dust lane on that spiral arm overlies the smaller galaxy, showing that NGC 5195 is somewhat further away along our line of sight. In a most intriguing paper (*Scientific American*, Dec. 1973, p. 38), Alar and Juri Toomre illustrated, by means of computer-simulation, the nature of the relationship – a 'hit-and-run' encounter in which mutual tidal influences are distorting both galaxies.

M51 is sufficiently nearby for the largest telescopes to resolve individual stars in its spiral arms. The amateur observer with a 20- to 25-cm instrument may, in conditions of good sky-transparency, detect something of the galaxy's spiral structure, and of the apparent bridge between the main body and the smaller NGC 5195.

M52
NGC 7654
Open cluster in Cassiopeia
RA   23h 23m

Dec +61° 29′

Mag   7.3

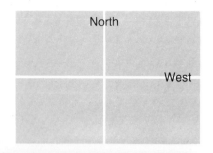

Visually, at low magnification, this object appears as a small nebulous wisp of light. At higher power it will be resolved into a tightly packed swarm of a few dozen faint stars. A fourteen-minute exposure taken through the 40-cm Newtonian reflector (above) reveals the cluster's hundreds of fainter members.

M52 is a young stellar family, similar in size and general appearance to the superb M37 in Auriga, but probably about 25 per cent further away. Among the stars of M52, the amateur observer usually detects chiefly the lucid bluish-white colour of B-type giants. The photograph, however, shows a hint of yellow in at least one of the most luminous members,

and a definite reddish hue in several fainter stars.

This fairly prominent cluster is a convenient guide for locating a more difficult object nearby: the delicate little nebula NGC 7635, illustrated a few pages on, will be found only about half a degree southwest of M52.

NGC 457
Open cluster in Cassiopeia
RA   01h 16m

Dec +58° 04′

Mag   7.5

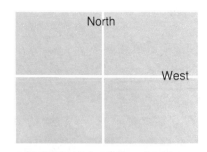

North

West

Many years ago the present writer, after observing this starfield with a wide-angle refractor, made the following notebook entry: '457 is a bright, roughly cruciform cluster, very dramatic at 20 ×. Distinctive yellow/blue pair of stars lie on the periphery.' Subsequent views of this cluster have always bourne out the initial favourable impression; it is a lovely object for the small telescope.

NGC 457 is a fairly dense aggregation of about a hundred luminous young stars within a sphere of space approximately two dozen light-years in diameter. There are indications that the dazzling yellowish star on the southeastern edge (φ Cassiopeiae) is actually a cluster-member, an F-class giant of enormous luminosity.

Just outside the field shown in the photograph, one degree northwest of NGC 457, lies a dimmer object, NGC 436, which appears visually as little more than a wisp of light. When viewed together in a wide field, at low magnification, NGC 457 and 436 are an interestingly contrasted duo; the fainter of the pair is a similar open cluster, situated in a considerably more distant region of space.

The Bubble Nebula
NGC 7635
Diffuse nebula in Cassiopeia

RA   23h 19m

Dec +60° 54′

Mag   8.5

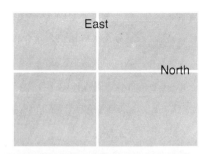

East

North

This small, faint object, described in the Webb Society's *Deep-Sky Observer's Handbook* as a wisp of material 'like a soap bubble in a steam cloud', is not easily observed visually. The precise nature of the Bubble Nebula has been a subject of divergent opinions; several writers have listed it as a planetary nebula.

A spectrosopic study by Italian astronomers Sabbadin and Bianchini (*Astronomy and Astrophysics*, vol. 155, pp. 177–82, 1977) has shown line-intensities in the nebula's spectrum that are uncharacteristic of either planetary nebulae or supernova remnants. The conclusion is that NGC 7635 is a diffuse cloud; the Italian colleagues identify the bright star within the nebula as its source of illumination. They also find evidence here of dense, compact globules, similar to the small obscuring masses within the better known interstellar gas clouds M8 and M16.

In the photograph we are probably viewing two separate objects. The fainter and more amorphous northern section of the nebulosity is designated as a discrete nebula (S162), perhaps more remote from us than the Bubble itself. The distances of both objects are rather uncertain.

The Perseus double cluster

NGC 869 and 884
Open clusters in Perseus

RA (NGC 869)   02h 16m (NGC 884)   02h 19m
Dec (NGC 869) +56° 55′ (NGC 884) + 56° 53′
Mags (NGC 869) 4.4   (NGC 884)   4.7

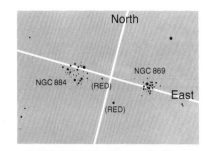

On a suitably clear, dark night an observer who directs his gaze at the region of the sky between Perseus and Cassiopeia will locate, without optical aid, a dim patch of light that gives the impression of a small, vague nebula. If he trains a telescope (preferably at low magnification) on this misty spot, he will discover the stunning twin object shown in the photograph. Each would be an outstanding cluster if it were alone; lying together within a single one-degree field,

NGC 869 and 884 comprise one of the most breathtaking small-telescope views in the sky.

Astrophysically, they are intriguing groups. Probably gravitationally connected, they both lie at a distance of somewhat more than 7000 light-years. Both are believed to be very young clusters of stars whose luminosities are enormous (the brightest 50 000 times greater than our Sun), but whose life-expectancies are short. The clusters, formed only a matter of some millions of years ago, did not begin their existence until the earth was already a mature, life-bearing world.

A prominent feature of the two groups is a sprinkling of especially bright stars whose red colour stands in dramatic contrast with the bluish-white of most of the cluster-members. These striking M-type red giants not only appear in the photograph opposite, but are also quite noticeable visually, through amateur telescopes of modest size.

M53
NGC 5024

RA   13h 12m
Dec +18° 17'
Mag   7.6

A distant globular cluster, some 65 000 light-years from us, M53 is still within the range that permits visual resolution of some member-stars with telescopes of 20 cm or larger.

NGC 4656

RA   12h 42m
Dec +32° 26'
Mag   11.0

This 11th-magnitude galaxy is classed as an 'irregular', similar to the famous Magellanic Clouds in its lack of apparent structure. It appears as a wispy streak of light in a 25 cm telescope.

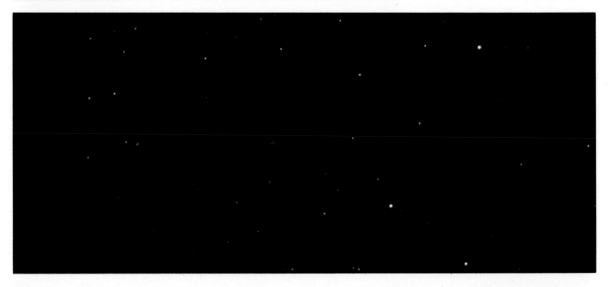

M56
NGC 6779
Globular cluster in Lyra
RA   19h 16m
Dec + 30° 08′
Mag   8.2

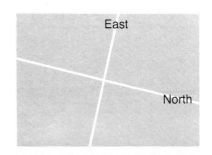

This little globular, which appears relatively faint and devoid of resolvable detail when viewed with a small telescope, is included as an instructive contrast with, for instance, the majestic Hercules cluster (M13) or the huge, sprawling M22 in Sagittarius.

On a summer evening it is a worthwhile exercise to view the two greater clusters first, and then to examine M56 at the same magnification. The dramatic difference of aspect, due in large part to the Lyra cluster's much greater distance from us, provides an insight into the three-dimensional depth of space. 45 000 light-years removed from our solar neighbourhood, M56 is nearly twice as far away as the better known Hercules cluster.

Rated as a somewhat sparse globular, M56 is intrinsically a rather less massive aggregation of stars than the very large clusters M13 and M22.

M57 The Ring Nebula

NGC 6720

Planetary nebula in Lyra

RA   18h 53m

Dec +33° 01′

Mag   9.3

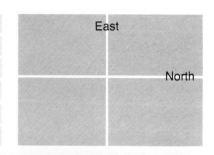

When telescopists of the 18th century began the first surveys of deep-sky objects, they stumbled across several minute oval spots of light that seemed to belong in a special class by themselves. Because their characteristic disc-shape was similar to the appearance of the more distant planets as viewed with the 6- or 8-cm telescopes of that era, astronomers called these the planetary nebulae.

M57, the 'Ring Nebula', is the best known of the class. Its popularity among amateur observers is due in part to its intriguing appearance in small instruments (the ring shape is revealed by even a 7-cm instrument), and in part to the

ease with which it can be located, for it lies directly on the short line between the stars β and γ Lyrae. A 15-cm telescope with a magnification of about 150 × shows the pale oval ring clearly differentiated from its dark central void. The central star shown in the photograph, however, is an elusive object in even fairly large telescopes, having a visual magnitude of only about 15, and perhaps at times less, if this star's suspected variability is a fact.

This star is the key to M57's nature. A blue dwarf of extremely high density and temperature, it is the end product of a collapse that followed (among other

evolutionary phases) the explosive creation of a surrounding envelope of gas. Given the nebula's probable diameter (about half a light-year) and rate of expansion, astrophysicists estimate that the explosive event may have occurred a little more than 20 000 years ago.

The nebula's distance is still a matter of uncertainty. M57 is not an exceedingly remote object; current measurements based on the dwarf star's probable intrinsic luminosity vary between about 1200 and 1500 light-years. At least one source, however (the Royal Astronomical Society of Canada's *Observer's Handbook*), places the Ring Nebula at 5000 light-years.

M58		
NGC 4579		
RA	12h 37m	
Dec	+11° 56′	
Mag	9.2	

This beautifully symmetrical face-on galaxy, classed as a barred spiral, is part of the Virgo Galaxy-Cluster. It lies immediately adjacent to M59.

M59		
NGC 4621		
RA	12h 41m	
Dec	+11° 47′	
Mag	9.6	

A giant elliptical galaxy, M59 is smaller but more massive than our Milky Way. At a distance from us of somewhat over 40 million light-years, M59 lies within the Virgo Cluster. At low magnification a small telescope shows both M58 and M59 within the same field.

M60

NGC 4649

Elliptical galaxy in Virgo

RA 12h 43m

Dec +11° 41′

Mag 8.9

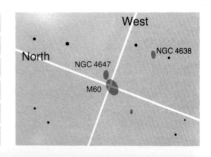

M60 is a galaxy very similar in many ways to M49, which lies only about 5 degrees to the southwest. Both are principal members of the Virgo Cluster, and both are unusually massive elliptical galaxies. M60 is five times as massive as our Milky Way.

The faint (12th-magnitude) galaxy that appears immediately northwest of M60 is not a background object, but actually a companion of the great elliptical. NGC 4647 is a small, loosely wound spiral galaxy. The rather ill-matched pair both lie at a distance of about 40 million light-years from us.

A small telescope provides a rather disappointing view of M60, which appears merely as a pale, hazy spot, through a typical 15- or 20-cm instrument. Some observers have managed to glimpse the elusive 12th-magnitude companion galaxy. By sweeping slightly more than one degree westward, one may locate another adjacent galaxy, the 11th-magnitude elliptical M59.

NGC 4567 and 4568
Spiral galaxies in Virgo
RA   12h 34m

Dec +11° 32'

Mag   12.0

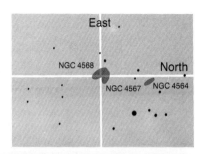

Located only about 2 degrees west of the fairly prominent Virgo galaxy M60, this fascinating pair of galaxies can be glimpsed visually with a moderate- to large-sized amateur telescope. Both of these spiral systems are somewhat more than 40 million light-years distant from us, and like M60 and other Messier galaxies in this part of the sky, they are members of the great Virgo Cluster.

At first glance, NGC 4567 and 4568 have the appearance of galaxies in collision or grazing interaction. It is interesting to consider the effects of such

encounters: many cases are known in which even a near miss between passing galaxies results in major tidal distortions of the two systems (cf. M51).

One might imagine that an actual head-on collision would be utterly catastrophic. In cases where we almost certainly are witnessing galactic collision (remote NGC 5128 in Cygnus is an example), extreme levels of radio emission are observed. Superheated by high-speed collision, the gaseous components of both galaxies are doubtlessly being scattered into intergalactic space, leaving the two systems so impoverished

that no further star formation will ever take place within them. Surprisingly, however, the passage of one galaxy directly through the body of another galaxy will probably result in no actual collisions between stars; as far as its stars are concerned, a typical galaxy is almost wholly 'empty' space.

Massive distortion of spiral arms and the more subtle phenomenon of filamentary bridges – the usual indications of tidal interaction between two galaxies – are absent in this pair. Very probably one is passing behind the other with a safe margin of space lying between.

M61

NGC 4303

Spiral galaxy in Virgo

RA   12h 21m

Dec +04° 36'

Mag   10.0

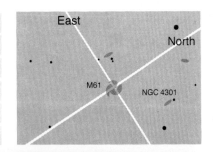

The photograph shows much detailed structure in this compact face-on system. Lying in the Virgo cluster of galaxies, at a distance of about 40 million light-years, it is a moderate-sized galaxy, of considerably smaller mass than our own.

The arrangement of the spiral arms in this galaxy's central region is rather ambiguous. Because of the oddly angular manner of the arms' attachment to the nucleus, some catalogues list M61 as a barred spiral.

Visual observation with amateur telescopes of average size reveals little detail in M61, which appears only as a pale smudge of light. The fifteen-minute exposure here, taken with a 40-cm reflector, includes images of several faint galaxies that lie in the remote distance, beyond M61.

M63
NGC 5055
Spiral galaxy in Canes Venatici
RA   13h 15m

Dec + 42° 08′

Mag   9.5

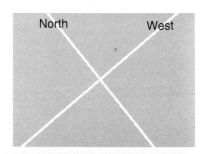

North   West

This spiral galaxy is roughly comparable in linear size to our own, and apparently nearly as massive. The structure of M63 is highly complex, with a tiny, button-like nucleus and an array of multiple arms that originate almost at the galactic centre. Numerous giant knots of luminous hydrogen gas (H II regions like the Lagoon and Orion nebulae in our own galaxy) lie strung along the full length of every spiral arm.

It is possible that this great star-system is associated with the M101 galaxy group; one recent estimate of its distance, however, places M63 about 35 million light-years away from us – twice the distance of M101.

Visually, M63 is not a difficult galaxy for small apertures. A 10-cm telescope shows it as a pale, elongated light-patch. With a 20-cm instrument, one can detect the sharp differentiation between this galaxy's bright core and its sparse, dim spiral arms.

M64 The Black-eye Galaxy
NGC 4826
Spiral galaxy in Coma Berenices

RA 12h 56m

Dec +21° 48′

Mag 8.8

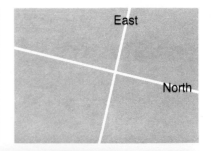

A favourite target for amateur galaxy-hunters, M64 is bright, compact and easily detected with a telescope as small as 6 cm. The extraordinarily extensive obscuring cloud of dust that borders this galaxy's nucleus is the feature that suggests the object's popular nickname. Although prominent, this dust lane is not easily seen visually with an instrument smaller than about 25 cm (yet, one expert observer has glimpsed it with a 10-cm refractor).

M64 is a tightly wound spiral system probably somewhat smaller than our own galaxy. Its distance has been estimated as 25 million light-years, but S. van den Bergh (1982) gives a figure that makes M64 a considerably nearer neighbour – 12 million light-years.

Many of the brightest galaxies that lie adjacent to M64 in the spring sky – M61, M84, M98, M100, etc. – form a cluster, all lying at an average distance of the order of 40 million light-years. M64 itself appears almost certainly not to be a member of the group, but a foreground object in intergalactic space.

M65 and M66
NGC 3623, 3627
Spiral galaxies in Leo

RA (M65)	11h 18m	(M66)	11h 19m
Dec (M65)	+13° 13′	(M66)	+13° 07′
Mags (M65)	9.3	(M66)	8.4

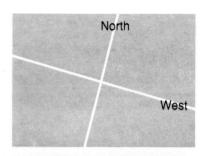

This close-knit pair of sister galaxies are well known to amateur astronomers because of the aesthetically pleasing aspect they present, lying together in the telescopic field. An instrument of 20- to 25-cm aperture or larger easily shows the fainter third galaxy in the group; the 10th-magnitude edge-on spiral NGC 3628 is only about half a degree north of the pair.

M65 and M66 are spiral galaxies of nearly equal size and general structure situated actually, and not just apparently, together in space at a distance from us of about 30 million light-years. Their somewhat different angles of orientation (M65 is the more nearly edgewise of the two) are surprisingly obvious when the two galaxies are viewed with even a small telescope. Both of these spirals are somewhat smaller than our Milky Way galaxy.

The neighbouring galaxy NGC 3628, just beyond the top edge of this field, appears on page 84.

NGC 3628	
RA	11h 8m
Dec	+13°53'
Mag	10.9

This large, edge-on galaxy appears in the same field, at low magnification, as the better known pair M65 and M66. At a distance of about 30 million light-years, NGC 3628 is closely associated in space with those Leo galaxies.

NGC 2903	
RA	09h 30m
Dec	+21°44'
Mag	9.9

Also located in the constellation Leo, NGC 2903 is a 10th-magnitude spiral system viewed in the face-on orientation. It is a complex, multi-armed galaxy whose bright nucleus can be spotted with a 15-cm telescope.

M71
NGC 6838
Globular cluster in Sagitta
RA   19h 53m
Dec +18° 44′
Mag   9.0

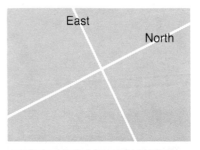

Because of its rather low surface brightness, this intriguing cluster is problematic for the user of a small telescope. Tantalizingly, at low power it just fails to be resolved into its component stars; yet it seems too faint to bear any higher magnification. Nevertheless, in the clean, unlit skies of rural Victorian England, T.W. Webb's 4-inch (10-cm) refractor showed M71 as a loose aggregation of small individual stars.

Thought by some of its early professional observers (Shapley among them) to be an open cluster relatively nearby, within our galaxy, M71 is now usually classed as a rather loose, uncompressed globular cluster. It lies at a great distance from us – about 18 000 light-years – and is about one quarter as large in overall diameter as the much more magnificent M13 in Hercules. Unlike that million-star system, however, M71 is a globular of fairly modest mass, lacking the densely crowded nucleus that seems to be a characteristic of the Hercules cluster and similar objects.

The wide-angle view provided by good astronomical binoculars shows M71 to good advantage, embedded in a spectacularly rich field.

NGC 7006
Globular cluster in Delphinus
RA   20h 59m
Dec +16° 00′
Mag   11.5

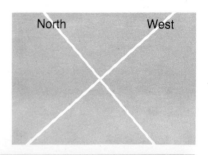

Invisible to small amateur instruments, this faint object is worth searching out with a 20- or 25-cm aperture, after first observing M13 or M15 with the same telescope. Like those glorious blazes of stars, this minuscule patch of light is a globular cluster.

While the impressive half-degree diameter of the great Hercules cluster fills the telescopic field of view, the same magnification shows NGC 7006 as an almost star-like spot in an expanse of empty space. The difference in appearance is the result of an almost staggering difference in the two clusters' intrinsic distances from us.

At a distance of 25 000 light-years, M13 is a globular lying in the fringe-region of our galaxy. NGC 7006 is enormously more remote – over 180,000 light-years distant. It has been suggested that this globular is a 'maverick', to be regarded as an intergalactic star-system, rather than as a component of the spherical halo of globular clusters that accompany our Milky Way galaxy.

Note that the portrait here, taken with a 40-cm telescope, resolves some of the individual 16th-magnitude stars that comprise this faraway denizen of the void.

M74
NGC 628
Spiral galaxy in Pisces
RA   01h 36m
Dec +15° 41′
Mag 10.2

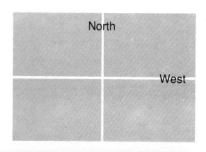

North

West

M74 is a geometrically pleasing spiral, viewed in the full-face orientation. An interesting feature is the relatively heavy bordering of this galaxy's arms by dust lanes that begin well within the central mass, and curve along the inner edges of the principal star-lanes.

Like M101, this system may be taken as a fair representation of our galaxy's probable appearance. Although similar to the Milky Way in form

and linear dimensions, however, M74 is a considerably less dense and thus less massive aggregation of stars than our own. The rather low surface-brightness that results from this galaxy's comparative sparseness makes it a dim object, visually. Experienced users of small telescopes consider this the most elusive of the Messier galaxies, and expect to capture a glimpse of it only in an unusually dark, transparent sky.

In a photometric study of selected regions in M74, Italian astronomers Guidoni, Messi and Natali observed a systematic reddening of the characteristic light, from the concave to the convex edges (*Astronomy and Astrophysics*, vol. 96, pp. 215–18, 1981). This phenomenon they interpret as evidence that the galaxy's direction of rotation is such that the convex edges are leading; that is, the spiral pattern is, as it were, being drawn around rather than pushed.

M76
NGC 650
Planetary nebula in Perseus
RA   01h 41m
Dec +51° 28′
Mag   12.2

North   West

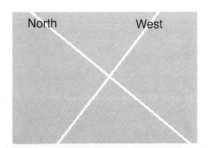

Beginning observers usually expect this tiny nebula (nominally the faintest of all the Messier objects) to be inordinately challenging. In fact, M76 comes as a pleasant surprise; its distinct, compact shape contrasts well with the sky-background, even if only a 12- or 15-cm telescope is used.

The *Webb Society Deep-Sky Observer's Handbook* describes the object's visual appearance thus: 'Hazy, nearly elliptical patch with waisted shape . . . Bar of faint light joining two sections with faint envelope surrounding . . .' These words might seem an appropriate description of the better known Dumbbell Nebula (M27) and, indeed, M76 is often called 'The Little Dumbbell'.

Like M27, this fainter object is a planetary nebula – a shell of gas expelled from a very hot central star, from which it derives its illumination.

The rather wide range of estimates that have been made of this planetary's distance suggest, variously, that it is a considerably smaller body than M27 and lying at about the same distance, or that it is similar in size to the Dumbbell, but further away. Planetary nebulae, as a class, continue to be among the most difficult objects to which to assign reliable distances; the true luminosities of their peculiar central stars are not easily estimated.

M77

NGC 1068

Spiral galaxy in Cetus

RA   02h 42m

Dec +00° 04'

Mag   8.9

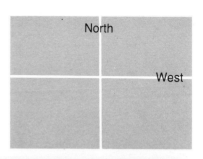

Although rather distant (about 60 million light-years), M77 is bright enough to be a relatively easy galaxy for the amateur to observe. A 10- or 15-cm glass shows its compact central hub and the outlying scatter of light that is a hint of spiral arms.

Oriented approximately face-on, as viewed from our own galaxy, this complex, multi-armed spiral is of a size, mass and superficial appearance not unlike our Milky Way system. In at least one respect, however, M77 exhibits the symptoms of peculiarity: strong radio emissions and other indications suggest that there is a violent eruptive event occurring in the galactic core. The system is thus a 'Seyfert galaxy', one of an intriguing class whose nuclei appear to be so energetically disrupted that they eject part of their mass into the outer parts of the affected galaxy, and into space beyond (cf. M87).

A moderate- or large-sized amateur telescope, sweeping the region within a radius of 2 degrees of M77, will detect several fainter galaxies immediately adjacent to the bright Cetus spiral. With M77, they comprise a loose group, or galaxy-cluster.

NGC 246
Planetary nebula in Cetus
RA   00h 45m
Dec − 12° 09′
Mag   8.5

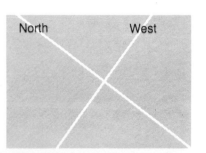

North   West

NGC 246 is a distended gaseous envelope surrounding a central, hot, white star; it is a classic planetary nebula comparable to the better known Ring Nebula (M57) and the Helix Nebula (NGC 7293). Like those objects, this gas-shell is the product of some sort of disruptive event in its central star. The impression that we are observing the debris of an actual explosion is probably not accurate; more likely NGC 246 and most of the similar planetaries are small masses of material lost from the star's superficial atmospheric layers by some less catastrophic process.

Uncertainty about the intrinsic luminosity of NGC 246's embedded central star has prevented any reliable estimate of the nebula's distance from us. (It is interesting to note that the distance of only one planetary nebula is known with certainty: the 14th-magnitude nebula surrounding a star in the globular cluster M15 is, of course, at the distance of that cluster.)

Visually, NGC 246 is an extremely challenging object. Like the Helix Nebula, this is a very large planetary, of low general surface brightness. A sketch by Canadian amateur P. Brennan (*Webb Society Deep-Sky Observer's Handbook*) shows that, viewed with a 20-cm telescope, the nebula appears as a faint, ghostly ring.

NGC 253
Spiral galaxy in Sculptor
RA   00h 47m

Dec −25° 34′

Mag   7.0

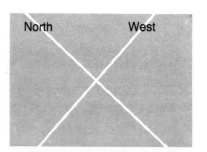

At a distance of just over 7 million light-years, this large spiral galaxy is a relatively close neighbour in intergalactic space – only about $3\frac{1}{2}$ times as far away as our great companion-galaxy M31 in Andromeda. NGC 253 is similar in size and structure to our Milky Way system, but viewed nearly edgewise.

The photograph taken with the 40-cm reflector shows much of the galaxy's intricate detail. Notice the compact central disc,

the two principal arms that sweep outward from points of attachment on opposite sides of the disc, and the mottled lanes of obscuring dust.

Because the plane of NGC 253 lies very nearly on our line of sight (its inclination is only about $11\frac{1}{2}$ degrees), its rotation is a dramatic motion as observed from our vantage-point: the outer portions of the north-eastern spiral arm are being swept toward us, while the extreme regions of the opposite

arm are being carried away from us. G. de Vaucouleurs has described the application of a device called the Fabry–Perot interferometer to measurement of this galaxy's rotation, in *Sky and Telescope*, Nov. 1981, p. 408. From detailed determination of the rates at which various parts of the galaxy rotate, the astronomer can estimate the total mass of the system. De Vaucouleurs gives a figure of 140 billion suns for the mass of NGC 253.

M78
NGC 2068
Diffuse nebula in Orion
RA   05h 46m
Dec +00° 02′
Mag   8.0

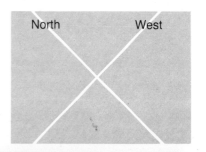

North   West

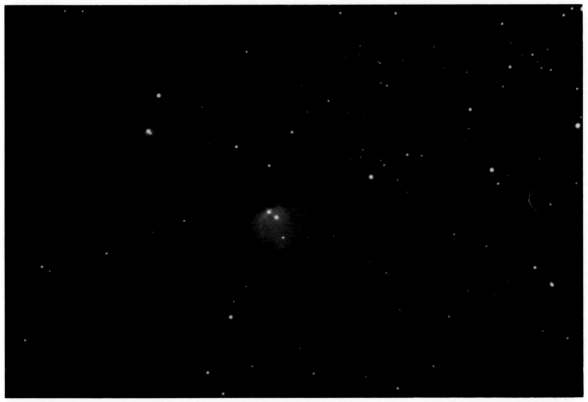

Often overlooked because of its proximity to the great Orion showpiece, M42, this fine little nebula is an easy and quite pleasing target for visual observation with small telescopes. Located just above the eastern end of Orion's belt, M78 appears through a 15-cm instrument as a widely spaced double star cushioned in a nest of pale mist.

Only a few light-years in overall diameter, the object is about the same size as a planetary nebula. It is a very different kind of object, however – a small surface-portion of a much vaster region of diffuse nebulosity, reflecting light cast upon it by hot foreground stars. It is, in fact, probably a separate outcropping of the huge gaseous cloud whose other visible lobes include M42 itself.

The pair of stars associated with M78 (both approximately of visual magnitude 10) actually lie within the nebulosity, rather than in the nearer foreground.

M81 and M82			
NGC 3031 and 3034			
Sb and Irr. galaxies in Ursa Major			
RA (M81)	09h 54m	RA (M82)	09h 54m
Dec (M81	+69° 09′	Dec (M82)	+69° 47′
Mags (M81)	7.9	(M82)	8.8

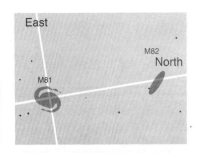

The photograph shows the one-degree telescopic field that includes the two great Ursa Major galaxies, M81 and M82, which rank among the most intriguing objects of their kind for the amateur observer. Lying close enough together to be seen simultaneously at low magnification, they present an instructive view of galaxies in the almost face-on (M81) and the almost edge-on positions. Both are sufficiently bright to be easily located with even a 7-cm telescope.

These two great stellar systems are situated in relatively close proximity to each other in space. They are, however, both well beyond the boundaries of our Local Group of galaxies, about five times as far away from us as our neighbour M31 in Andromeda. In spite of this great distance, the largest optical telescopes resolve bright H II regions and some of the most luminous of the galaxy's individual stars in M81's magnificent spiral arms.

M82 presents a more enigmatic picture. Not only does this cigar-shaped galaxy permit no resolution of individual stars but, more strangely, its long-exposure portrait includes radiating filaments that look very much like a catastrophic splash of material flying outward from the central regions. Most astronomers have seen this galaxy's intense radio emission and strong light-polarization as corroboration of the impression given by its photographic image: that M82 is an exploding galaxy.

The explosion theory, however, is not altogether uncontroversial. Philip Morrison (*Sky and Telescope*, Jan. 1979, pp. 26–31) has suggested a strikingly different model consistent with the same observed data: a healthy, undistressed galaxy drifting through a relatively static envelope of dust. This intergalactic cloud may be material drawn out of both M81 and M82 – perhaps a fairly normal product of the tidal interaction between two neighbouring galaxies.

M83
NGC 5236

RA	13h 36m
Dec	−29° 46′
Mag	10.1

M83 is both an intrinsically large and relatively nearby spiral galaxy, at a distance of only about 10 million light-years. It is regarded as especially worthy of close photographic monitoring, because of the abnormal frequency of supernova-outbursts in this galaxy.

M85
NGC 4382

RA	12h 24m
Dec	+18° 18′
Mag	9.3

This relatively bright and easily detected Virgo Cluster member is usually classed as an 'SO' or transitional type of galaxy. The little spiral just to the east of M85 is 12th-magnitude NGC 4394.

M84 and M86
NGC 4374 and 4406
Elliptical galaxies in Virgo

RA (M84)	12h 24m	(M86)	12h 25m
Dec (M84)	+13° 00′	(M86)	+13° 03′
Mags (M84)	9.7	(M86)	9.3

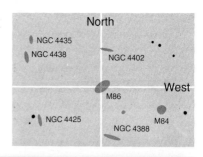

These large elliptical galaxies in Virgo have an apparent separation of only about 17 arc-minutes, giving the impression of two very nearly equivalent objects that lie close together in space. They are, in fact, rather dissimilar in size and one of the pair is greatly more distant from us than the other.

M84 (the more westerly of the two) is a member of the Virgo Cluster of galaxies, at a distance of the order of 40 million light-years. It is a giant galaxy of some 500 billion solar masses, with a probable diameter so great that light takes 130 000 years to cross it.

M86 is a considerably smaller object at perhaps half the distance of its companion. Its maverick motion relative to galaxies of the Virgo Cluster – actually 450 kilometres per second *toward* our position in space – may indicate that M86 is a galaxy having no connection with the cluster.

The two ellipticals (magnitudes 9.3 and 9.7) can be observed visually with small amateur telescopes. A 12.5-cm Schmidt-Cassegrain, for instance, reveals them as a close pair of rather faint light-smudges, if conditions of seeing are favourable.

Other galaxies in the same field include the slender edge-on spirals NGC 4388 (magnitude 11) and NGC 4402 (mag 11.5). More easily seen are the very closely spaced pair NGC 4438 (mag. 11.0) (spiral) and NGC 4435 (mag 11.8) (elliptical), which lie directly east of M84 and M86.

M87
NGC 4486
Elliptical galaxy in Virgo
RA   12h 30m
Dec + 12° 30′
Mag   9.2

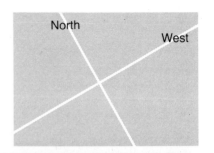

Listed in many catalogues as a 'peculiar' galaxy, M87 is extraordinary in most of its aspects. It is an armless, spherical galaxy like M32 (the small companion to the Andromeda system), but on an awesomely gigantic scale; its stellar population is perhaps thirty times that of our Milky Way.

Although its visual and photographic appearance is rather featureless, M87 is an unusually interesting object. An enormously strong source of radio and X-ray emissions, it is clearly a galaxy that is experiencing some manner of violent internal upheaval. Photographs made with the world's largest telescopes reveal a more dramatic indication, in the form of a long jet of material that appears to be hurtling outward from the galactic nucleus.

Close inspection of the illustration above will bring another feature to light: the small, fuzzy spots that fringe the galaxy's perimeter are not stars individually resolved, but globular clusters similar to M13 in our own galaxy. M87 is attended by a vast number of these satellite systems – probably several thousands of them.

This intriguing galaxy lies at a distance of about 40 million light-years, in a cluster of galaxies that includes M61, M84, M100 and many others that can be detected with ordinary amateur telescopes.

M88
NGC 4501
Spiral galaxy in Coma Berenices
RA   12h 31m

Dec +14° 32′

Mag   10.2

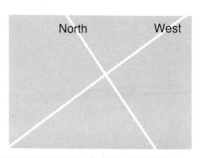

Situated about 41 million light-years distant from us, this spiral is apparently another member of the sprawling galaxy-cluster in Virgo and Coma Berenices.

M88 is a large system of highly regular shape, with a bright central bulge and well developed, symmetrical arms. Vehrenberg (*Atlas of Deep-Sky Splendours*) notes the similarity between this great spiral galaxy and the much nearer and more familiar M31 in Andromeda.

M88 is not well known to most amateur observers, perhaps because it is omitted in many of the classic handbooks, including Webb's *Celestial Objects for Common Telescopes*. Although its nominal magnitude of 10.2 sounds forbiddingly faint, the galaxy is a surprisingly easy one for modest apertures, being a compact object with fairly intense nucleus. In a 15-cm telescope at low magnification, it is a small glowing patch whose elongation in a northwest–southwest direction is readily perceived.

M89
NGC 4552
RA 12h 35m
Dec +12° 40′
Mag 9.5

This is yet another of the giant, very massive, elliptical galaxies that abound in the Virgo Galaxy-Cluster. Although visually compact and featureless, it is a dense throng of 250 billion solar masses.

NGC 4559
RA 12h 34m
Dec +28° 14′
Mag 10.6

Located due north of M89, in Coma Berenices, this multi-armed spiral is a considerably more challenging target for small telescopes.

M90	
NGC 4569	
RA	12h 36m
Dec	+13° 16′
Mag	10.0

Lying adjacent to M89, about a degree to the northeast, M90 is a spiral galaxy whose pale, 10th-magnitude glow is near the limit of perception in a 10-cm telescope. A notable feature of this system is its bright, highly condensed nucleus.

This loosely structured spiral is a much more difficult object than M90. Although of nominally similar magnitude, it is a large system having a low general surface brightness; visually, it is barely perceptible with even a 40-cm aperture.

NGC 4395	
RA	12h 23m
Dec	+33° 49′
Mag	11.4

M92
NGC 6341
Globular cluster in Hercules

RA   17h 17m
Dec +43° 10′
Mag   6.1

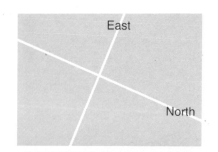

The summer constellation of Hercules is blessed with not merely one, but two, of the most magnificent globular clusters visible from northern latitudes. Less than half a magnitude fainter than M13, and of approximately the same impressive apparent size, M92 is a major globular cluster lying only about 25 per cent further away from us in space.

This second great Hercules cluster is a massive aggregation of not less than several hundred thousand stars, with the same extraordinary degree of compression and high density in its blazing core-region observed in M13. A comparison of the photograph with that of the other globular shows the somewhat looser and more straggling sprawl of outlying stars that characterizes M92. Intriguingly, one star of this cluster's teeming membership has been identified as an eclipsing binary (an orbiting pair, one component of which eclipses the other, as viewed from our vantage-point on the earth).

Several classic observers, including Webb and Buffham, rated M92 much below M13 in visual splendour, and described the slightly more distant cluster as a considerably more difficult telescopic object, 'resolved in glimpses'. To the present writer, however, M92 is equally attractive small-telescope fare. Viewed through a 12-cm refractor, it displays a brilliant core and extensive surrounding regions whose most luminous individual stars (visual magnitude about 12.5) are visible as tiny points of light.

M94

NGC 4736

Spiral galaxy in Canes Venatici

RA   12h 50m

Dec +41° 14′

Mag   7.9

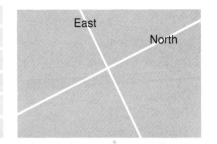

With a stellar population equivalent to 250 billion suns, M94 is a galaxy as massive as our own. It is structurally unlike the Milky Way, however, in the unusual tightness of its spiral arms; they are scarcely divided from each other, or from the abnormally large, featureless core-region.

An intriguing feature has been detected in very long photographic exposure of this system: a faint elliptical ring of exceedingly tenuous material surrounds the entire galaxy, well beyond its apparent outer perimeter. There is evidence of a further peculiarity in M94, in that its central bulge and the outer disc appear to have different planes of rotation.

Visually, this galaxy is a joy to observers having only modest equipment. At a distance of about 14 million light-years, it is sufficiently bright to be spotted easily with even a 6-cm refractor.

NGC 4631
Spiral galaxy in Canes Venatici
RA   12h 40m

Dec +32° 49′

Mag   9.3

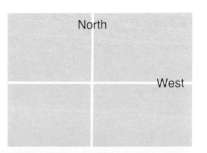

North

West

Oriented so precisely edgewise that identification of its structural pattern is unclear, NGC 4631 is regarded as probably a loosely wound spiral whose appearance, if it could be viewed in full face, would not be unlike that of M74.

This galaxy has an unusual lack of the heavy, continuous dust lane that often bisects spiral galaxies viewed in this orientation. Photographs by even the great 5-m Hale reflector reveal only small, separate patches of obscuring material along the equatorial plane.

The small object immediately adjacent, northwest of the central bulge of NGC 4631 is the dwarf elliptical system NGC 4627. It is believed that this little galaxy forms a dynamical unit with its huge neighbour, just as M32 and NGC 205 are satellites of the Great Andromeda Galaxy.

Another stellar system in this part of the sky, the irregular galaxy NGC 4656, lies just half a degree southeast of 4631. On long exposures encompassing both objects there is evidence of a tenuous 'bridge' of gaseous material spanning the void between the two galaxies.

For small-telescope observation, the great edge-on spiral is not a difficult object, appearing as a slender streak of light in which, with an instrument of 20 cm, some condensations and mottling may be seen.

M96	
NGC 3368	
RA	10h 46m
Dec	+11° 56′
Mag	9.1

This is a spiral galaxy, with arms so tightly wound that the overall appearance is that of a virtually featureless ellipse. M96 is probably associated in space with the galaxy-group that includes M65/66.

M105	
NGC 3379	
RA	10h 47m
Dec	+12° 42′
Mag	9.2

Situated within the same low-power field with M96, in a rich-field telescope, M105 is an elliptical galaxy, actually associated in space with its apparent neighbour.
(Photo enlarged above standard scale)

M97 The Owl Nebula
NGC 3587
Planetary nebula in Ursa Major

RA   11h 14m

Dec +55° 08′

Mag   12.0

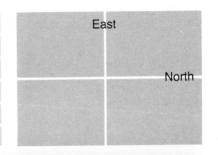

East

North

M97 is the spheroidal shell of a gas expelled from a star whose very hot, dwarf remnant is the luminous point visible at the nebula's centre. Although generally comparable to the noted Ring Nebula (M57) and the Dumbbell (M27), M97 is visually a much dimmer object. Viewed with a 10- or 12-cm telescope, it is a barely perceptible, and quite featureless, circular patch of haze.

The photograph reveals this planetary's intriguing inner structure. The two dark voids, which were suggestive to Lord Rosse of a pair of round eyes, prompted that 19th-century observer to call this nebula the Owl.

Like several of the other planetaries, M97 poses continuing difficulties in connection with the estimates of its distance and true size. The distance figures quoted in various handbooks range from 1600 to over 10000 light-years; thus, the object's intrinsic diameter may be as little as two, or as large as perhaps nine light-years. Recent studies have shown increasing agreement on a distance in the smaller half of the suggested range.

It is interesting to bear in mind, while looking at the photograph opposite, that this large and seemingly massive object is, in fact, a highly rarified bubble containing a small fraction of one solar mass, in total.

M98
NGC 4192
Spiral galaxy in Coma Berenices
RA   12h 13m
Dec +15° 01′
Mag 10.7

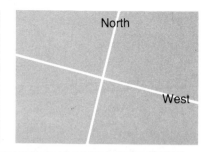

M98 is a large and massive spiral galaxy viewed almost edgewise. A notable structural feature of this great star-system is its very small, highly compressed core and, by contrast, sparse outer regions.

Its estimated distance of about 35 million light-years makes M98 a probable member of the galaxy-cluster in Virgo and Coma Berenices, with which M60, M88, M100 and many others are associated.

The present writer has found M98 to be one of the most difficult of the Messier galaxies to observe with a telescope of 10- to 15-cm aperture. Although considerable help in locating the rather dim galaxy is provided by its position just half a degree west of the star 6 Leonis, it can be missed if sky-conditions are less than ideal. On a night that is free of haze and moonlight, M98 appears visually as a subtle wisp of light, elongated in the northwest–southwest direction.

M99
NGC 4254
RA   12h 18m
Dec +14° 32′
Mag   10.1

Viewed in the face-on orientation, this beautifully symmetrical spiral lies only a little more than one degree eastward of M98. It is an easier object than that galaxy, when observed with a telescope of 10- to 15-cm aperture.

M102
NGC 5866
RA   15h 06m
Dec +55° 50′
Mag   10.8

Appearing to the visual observer as a very pale sliver of nebulosity, M102 is a highly elongated elliptical galaxy. A 20-cm telescope gives an impression very close to that captured here on film.

M100
NGC 4321
Spiral galaxy in Coma Berenices
RA   12h 22m

Dec + 15° 56′

Mag   10.6

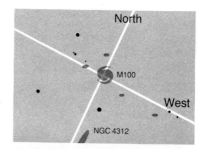

Its two principal arms are this galaxy's major feature, each being, on the average, about 2900 light-years in breadth – twice the average thickness of typical spiral arms in our own galaxy. Although M100 is not a close neighbour in intergalactic space (its distance from us is of the order of 40 million light-years), it is still within the range that permits resolution of individual stars in the best photographs by the world's largest telescopes.

In the portrait of M100 taken with the 40-cm reflector, a bright edgewise galaxy (NGC 4312) is readily seen to the southeast of the great spiral. Closer inspection of the picture will reveal a half dozen smaller and much fainter objects surrounding M100. These are dwarf elliptical galaxies similar in size and in their relationship with the massive principal galaxy to the more familiar Andromeda system's two well known companions M32 and NGC 205.

At visual magnitude 10.6, M100 is a relatively undistinguished galaxy for the small telescope.

NGC 4565
Spiral galaxy in Coma Berenices
RA   12h 34m
Dec +26° 16′
Mag   10.2

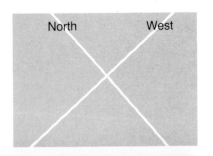

North   West

A classic edge-on spiral, NGC 4565 presents the finest possible view of a galactic profile; the photograph here shows the typical central bulge and flattened disc of a moderately loose-armed spiral, with a thin, dense lane of obscuring dust clearly delineating the equatorial plane.

NGC 4565 is usually regarded as having an especially instructive appearance. When, from the earth's position in the outer regions of our own galaxy, we view the Milky Way lying in a bright, narrow band across the night sky, we are observing a system like NGC 4565 at point-blank range. If we could withdraw to a vantage-point about 20 million light-years more distant from the Milky Way, it would shrink to an aspect like that of NGC 4565.

Although this fine Coma object is an intrinsically large and relatively nearby galaxy, its orientation results in little of its light being directed earthward; it is thus a dim and elusive sight in a small instrument. Nevertheless, a telescope of 15-cm aperture shows NGC 4565 as an intriguing sliver of light in a suitably dark, transparent sky. Because its position adjacent to the bright Coma star cluster (Mel 111) renders it easy to locate, it is a favourite target for amateur galaxy-hunters.

M101

NGC 5457

Spiral galaxy in Ursa Major

RA   14h 02m

Dec +54° 27′

Mag   9.6

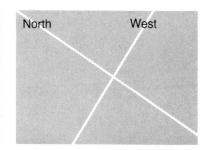

North   West

The visual appearance of M101 with an amateur's telescope is unimpressive; one sees only a pale, elusive patch of light scarcely distinguishable from the general sky-glow.

The long-exposure photograph shows this object as a loose-armed (Sc) spiral galaxy viewed directly face-on – one of the most magnificent of its kind. *The Hubble Atlas of Galaxies* describes this as the prototype of the 'multiple arm' galaxies, with the outer arms heavily branched. Characteristic features seen in the photograph include small clumps of luminous nebulosity and black obscuring dust lanes. The elongated bright feature at the end of the eastern arm is a vast star-cloud, measuring approximately 2000 light-years on its major axis.

Although almost as great in diameter as our own Milky Way system, M101 is a relative featherweight among spiral galaxies; it contains material that totals only about 8 per cent of our galaxy's mass.

At a distance of 14 million light-years, M101 lies beyond the boundaries of our Local Group of galaxies. It is, in fact, one of about ten systems that comprise a neighbouring group having a mixed population of large spirals, dwarf ellipticals and one irregular galaxy – a galactic family closely analagous to our own. Another major spiral of the group (NGC 5474) is less than a degree southeast of M101, just beyond the frame-edge of our photograph. The much fainter galaxy visible near the northeastern corner of the field is probably a distant object in the remote background, rather than a member of the M101 group.

NGC 5907
Spiral galaxy in Draco
RA   15h 15m
Dec   +56° 31′
Mag   11.3

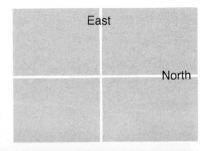

The visual magnitude of this object is quoted with wide variations in various catalogues. The *Webb Society Deep-Sky Observer's Handbook* estimates the galaxy's magnitude as 10.4 and describes NGC 5907 as 'bright and very elongated, with a brighter middle; no sign of absorption lane'. This is a visual impression of the edge-on spiral's appearance through a large amateur instrument. It can probably be glimpsed with an aperture as small as 12 cm.

NGC 5907 is a loose-armed system whose general layout, if it could be viewed face-on, would generally resemble the famous Sc galaxy M101.

An edgewise orientation in galaxies is useful for studies that attempt to establish rotational direction; the opposite ends of such a 'spindle' galaxy show a maximum difference in shifts of spectral lines, caused by the fact that the arms on one side of the central bulge are directly receding from us, and arms on the opposite side are directly approaching.

Spectroscopic observations of NGC 5907 indicate that the south end of the spindle is the approaching edge of the galaxy, while the northern part is receding.

M103
NGC 581
Open cluster in Cassiopeia
RA   01h 32m
Dec +60° 35′
Mag   7.4

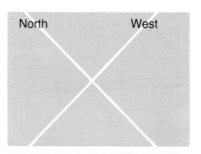

So loose and sparse is this little galactic cluster that many beginning observers fail to recognize it as a cluster at all. Although Shapley once expressed doubts that M103 is a true physical association, the group is now understood to be a gravitationally related family of about 40 identifiable members. Kenneth Glyn Jones has noted that, with an overall diameter of 15 light-years, M103 has a density of 5.76 stars per cubic parsec in the region of its core.

The comparative brilliance of the cluster's principal half-dozen stars, in spite of their considerable remoteness (about 8500 light-years from us), gives a clue to their true luminosity. They are giants displaying the characteristic, flashing blue-white colour of spectral type B. Among the cluster-stars recorded in the photograph here, hints of other colours can be made out, including the dull reddish hue of at least one M-class giant.

The distinctive triangular shape of M103 is easily recognized at low magnification. Using higher telescopic power, one may resolve Σ131, at the cluster's northern apex. This fine double star, with components of contrasting yellowish and blue colours, can be divided with a good 6-cm instrument.

NGC 663

RA   01h 43m
Dec +61° 01′
Mag   7.1

It is worthwhile, after observing the well known Cassiopeia cluster M103, to sweep 3 degrees northeastward to locate NGC 663, a subtle patch of 7th magnitude stars. This little open cluster includes about eighty members.

NGC 6543

RA   17h 59m
Dec +66° 38′
Mag   9.0

This tiny planetary nebula in Draco, easily seen as a fuzzy, greenish disc through a 10-cm telescope at high magnification, is only about one-third of a light-year in overall diameter. The small size of this gaseous shell is evidence that its material was expelled from a central star quite recently.

M104 The Sombrero Galaxy
NGC 4594
Spiral galaxy in Virgo

RA   12h 39m

Dec − 11° 31′

Mag   8.7

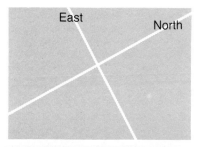

This is a galaxy viewed nearly edge-on; its disc-plane is inclined only 6 degrees to our line of sight. M104, noted for its unusually prominent equatorial lane of obscuring dust, is one of several edge-viewed systems (cf. also NGC 4565) whose appearance suggests that most of the dust in galaxies tends to be confined to a thin plane, perpendicular to the axis of rotation.

If M104 could be observed face-on, it would be seen as a fairly tight, very regular spiral, with arms wound in a smooth circular fashion. With a diameter of 140 000 light-years and a mass $6\frac{1}{2}$ times that of our Milky Way system, the 'Sombrero' is a giant galaxy. At a distance of over 37 million light-years, it is not a close neighbour in intergalactic space.

A series of radial velocity measurements of individual regions of M104 by Pease has yielded an odd result: this spiral appears to be rotating in the 'unwinding' direction (that is, with the open ends of the arms leading), unlike other galaxies that have been studied in this way.

M106

NGC 4258

Spiral galaxy in Canes Venatici

RA   12h 18m

Dec +47° 25'

Mag   8.6

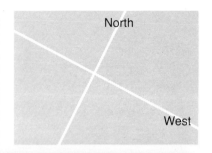

North

West

Here is one of the handful of galaxies classed by amateur observers as really easy objects to observe with a small instrument. With a 12-cm Schmidt–Cassegrain telescope, the present writer has found it a bright, compact smudge of light, distinctly elongated on a roughly north–south axis.

M106 is a spiral with moderately tightly wound arms. This intrinsically large and very massive system, at only about 14 million light-years distance, is a relatively close neighbour to our own galaxy. Some photographs by the world's largest telescopes reveal indications of peculiar activity in the nucleus of M106,

with evidence of material being ejected. The galaxy is a source of radio emission.

The edge-on spiral galaxy NGC 4217, whose subtle spindle-shape can be glimpsed with a 20-cm aperture, lies only about half a degree southwest of M106.

NGC 4449
Irr. galaxy in Canes Venatici
RA   12h 26m
Dec +44° 22′
Mag   9.2

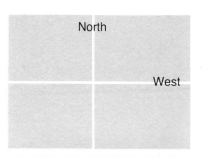

North

West

The majority of galaxies (and virtually all of the galaxies illustrated in this book) fall into the morphological groups described by astronomers as the spiral, barred spiral and elliptical types.

Some systems are observed, however, that do not fit into any of these principal categories. These atypical galaxies, characterized by an amorphous structure devoid of arms or rotational symmetry, and often by their dwarfishness in mass and linear size, are labelled 'irregular'. Our Milky Way galaxy's nearby satellites, the two Magellanic Clouds, are the best known examples.

In the irregular galaxy whose portrait appears here we are able to note the characteristic armless, rather angular sprawl of stars and gas. NGC 4449 is typical of its class in the large and numerous H II regions (concentrated chiefly in the northern regions, and readily visible in the photograph) and the discrete patches of dust that cluster near the galactic core.

NGC 4490
Spiral galaxy in Canes Venatici
RA   12h 28m

Dec +41° 55′

Mag   9.7

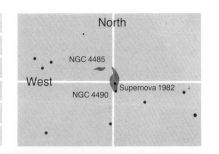

This bright galaxy can be glimpsed with a 10-cm telescope, and begins to show some hint of structure when observed with 20-cm and larger instruments. It is a loose spiral, a classic of the twin-armed, S-shaped type. The galaxy NGC 4485, immediately adjacent (northwest), is believed to be a near companion of NGC 4490 in space.

As it happens, the photograph records a most unusual moment in the galaxy's history. The reddish star that appears just south of the galactic core was the centre of a dramatic series of events in the spring of 1982.

Jack Newton obtained the photograph on the evening of March 29, 1982, while routinely surveying a number of galaxies, in search of supernovae. When the sudden onset of clouds prevented him from completing the roll of film, he decided to leave the film in the camera for use on a subsequent night.

On April 21, when this almost forgotten film was at last processed, Newton immediately noticed the star (not seen in earlier portraits of NGC 4490) near the central bulge. Upon telephoning the Dominion Astrophysical Observatory near his home in Victoria, British Columbia, he was informed that a telegram had been received at the observatory, reporting an observation of the supernova by Paul Wild. Ironically, Wild's discovery of the stellar explosion in NGC 4490 had occurred nearly seventeen days after the night of Newton's photograph – but nearly a week before the actual processing of the film. In a state of profound chagrin, Newton resolved never again to let exposed film lie undeveloped inside a camera.

Fuller details about the NGC 4490 supernova can be read in *Sky and Telescope* vol. 63 (July 1982), p. 106.

M107
NGC 6171

RA   16h 31m
Dec −13° 02′
Mag   9.2

Located in the constellation Ophiuchus, this small, distant globular cluster contrasts interestingly with major specimens of the class, such as M13 and M22.

M108
NGC 3556

RA   11h 11m
Dec +55° 47′
Mag 10.7

The dim, obliquely oriented spiral galaxy M108 lies about a degree north of the Owl Nebula (M97). Although very faint in a 10- or 15-cm telescope, the galaxy reveals its distinct east–west elongation.

M109

NGC 3992

Barred spiral in Ursa Major

RA 11h 57m

Dec +53° 29'

Mag 10.8

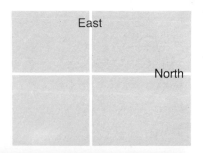

East

North

Although definitely not one of the easier Messier galaxies for visual observation with a small telescope, M109 is revealed on film as a superb barred spiral. It is characterized by a small nucleus, a broad transverse bar and thin, loosely wound arms.

The barred spirals, which comprise a minority among galactic types, were long believed to be a structurally distinct species; it was supposed that they were products of an evolutionary process separate from that of galaxies displaying the more common spiral form. Recent high-resolution photographic surveys have shown that the two galactic types are not really distinct. Numerous examples are found of simple spirals that, on close inspection, show vestiges of a central bar.

Viewed through a telescope of 10- or 15-cm aperture, M109 is a rather disappointing galaxy, appearing as little more than a pale, featureless glow at the very limit of visual perception.

M110

NGC 205

Elliptical galaxy in Andromeda

RA   00h 39m

Dec +41° 35′

Mag   9.4

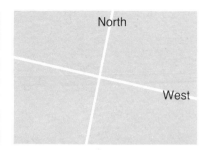

North

West

Located only about half a degree from the centre of the noted Andromeda Galaxy (M31), this much smaller and fainter object can be glimpsed in the same telescopic field at low magnification. Through a 10-cm aperture, it appears as a feeble smudge of light, elongated on a southeast–northwest axis. Little more detail is seen with amateur instruments of even considerably larger size.

This dwarf elliptical galaxy, a satellite of its vastly more massive neighbour M31, is an exceedingly interesting little system. Because of its relative proximity to our own galaxy, NGC 205 can be resolved into individual stars on photographs

made with very large telescopes such as the great Hale 5-m reflector. Concerning photography of NGC 205, Sandage (*The Hubble Atlas of Galaxies*) points out an odd fact: since all the stars resolved in NGC 205 are of about the same apparent brightness, 'resolution does not occur until a critical exposure time is reached, at which time the entire smooth edge of the galaxy breaks up into individual stars'.

Unlike most 'pure' elliptical systems, NGC 205 contains some prominent dust regions and also a suspected envelope that may extend well beyond the visible periphery of the galaxy – a feature that suggests some sort

of transitional status (cf. NGC 404, an 'SO' system, characterized by its distended envelope).

Two researchers (Gallagher and Mould, in *Astrophysical Journal*, vol. 244, pp. 13–16, 1981) have identified a feature that 'opens some interesting possibilities for . . . the evolutionary history of this curious Local Group member'. Their discovery of luminous young red stars adjacent to the galaxy's dust-clouds may be evidence of a unique burst of fairly recent star formation – something not expected in the elderly stellar population of a typical dwarf elliptical galaxy.

DATA FOR OBSERVERS

These tables of data provide a concise survey of objects described or mentioned in the preceding section. Abbreviations under the 'Class' heading are: D.Neb. – diffuse nebula; E.G. – elliptical galaxy; G. – galaxy; G.C. – globular cluster; G.Cl. – galaxy cluster; G.Irr. – irregular galaxy; G.SO – 'transitional' galaxy; O.C. – open cluster; O.Neb. – obscuring nebula; P.Neb. – planetary nebula; S.G. – spiral galaxy.

'Tel. Aperture' is a very rough guide (in cm of objective-diameter) to the smallest telescope that may be expected to give a satisfactory view of each object. Although this is necessarily rather subjective, it will serve to identify those objects most likely to be impressive with instruments of, say, 6- or 7-cm aperture, and those that certainly require a greater light-grasp for minimal satisfaction.

Object	Page	RA h m	Dec ° ′	Class	Mag	Tel. Aperture	Comments
M1	11	05 33	22 01	P.Neb.	8.4	6 cm	Small, pale smudge with most apertures
M2	12	21 32	−00 54	G.C.	6.3	6 cm	Needs 20 cm for resolution of individual stars
M3	13	13 41	28 29	G.C.	6.4	6 cm	Resolved with 15 cm
M4	14	16 22	−26 27	G.C.	6.4	6 cm	Large, loose cluster; rather dim
M5	15	15 18	02 10	G.C.	6.2	6 cm	Some resolution with 10 cm
M8	16	18 02	−24 23	D.Neb.	6.0	6 cm	6 cm shows cluster, hint of nebulosity
M10	17	16 56	−04 05	G.C.	6.7	8 cm	Needs 20 cm to begin resolution
M11	18	18 50	−06 18	O.C.	6.3	6 cm	6 cm shows swarm of tiny stars
M12	17	16 46	−01 55	G.C.	6.6	8 cm	Fainter, looser than M10
M13	22	16 41	36 30	G.C.	5.7	6 cm	Considerable resolution with 10 cm
M15	24	21 29	12 05	G.C.	6.0	6 cm	Slightly more difficult than M13
M16	29	18 18	−13 48	D.Neb.	6.5	20 cm	Smaller aperture shows cluster, no nebulosity
M17	30	18 20	−16 12	D.Neb.	7.0	6 cm	One of the brightest nebulae for 6 cm
M20	31	18 01	−23 02	D.Neb.	9.0	10 cm	Rifts glimpsed with 20 cm
M22	32	18 35	−23 56	G.C.	5.9	6 cm	Very easily resolved with 12–15 cm
M27	33	19 59	22 40	P.Neb.	7.6	6 cm	Elongated shape clear with small apertures
M29	34	20 23	38 27	O.C.	7.1	6 cm	Interesting at low power
M30	39	21 39	−23 15	G.C.	8.4	15 cm	Small, disappointing globular
M31	40	00 42	41 09	S.G.	4.8	5 cm	Impressive binocular object!
M32	41	00 42	40 45	E.G.	8.7	8 cm	In field just S of M31
M33	47	01 33	30 33	S.G.	5.8	8 cm	Pale object, needing low magnification
M34	48	02 41	42 43	O.C.	5.5	5 cm	Bright, loose cluster; use low magnification
M35	49	06 08	24 21	O.C.	5.3	6 cm	15 cm shows NGC 2158, adjacent
M36	51	05 35	34 05	O.C.	6.3	6 cm	Bright, loose cluster
M37	52	05 52	32 33	O.C.	6.2	8 cm	Condensed mass of tiny stars
M38	53	05 27	35 48	O.C.	7.4	8 cm	15 cm shows NGC 1907, adjacent

Object	Page	RA h m	Dec ° ′	Class	Mag	Tel. Aperture	Comments
M39	56	21 32	48 21	O.C.	5.2	5 cm	Very large; needs low magnification
M42	57	05 34	−05 24	D.Neb.	4.0	5 cm	Finest object for all apertures!
M43	57	05 35	−05 18	D.Neb.	8.0	10 cm	Faint circular glow N of θ_1 Orionis
M44	63	08 39	20 04	O.C.	3.7	5 cm	Best at about 15–20 $\times$
M45	64	03 46	24 03	O.C.	1.4	eye	Perhaps most dramatic with unaided eye
M46	65	07 41	−14 46	O.C.	6.0	6 cm	15 cm shows tiny nebula on N edge
M47	66	07 36	−14 27	O.C.	4.5	5 cm	Good binocular object
M49	67	12 29	08 07	E.G.	8.6	8 cm	A featureless glow, with all apertures
M51	69	13 29	47 18	S.G.	8.1	8 cm	30 cm shows hint of spiral structure
M52	70	23 23	61 29	O.C.	7.3	6 cm	Just resolved by this aperture
M53	74	13 12	18 17	G.C.	7.6	10 cm	Small, dim; resolved by 30 cm
M56	75	19 16	30 08	G.C.	8.2	10 cm	Like M53; slightly looser
M57	76	18 53	33 01	P.Neb.	9.3	10 cm	Ring shape glimpsed with 6 cm
M58	77	12 37	11 56	S.G.	9.2	10 cm	Visually dim, featureless
M59	77	12 41	11 47	E.G.	9.6	10 cm	In same field with M58
M60	78	12 43	11 41	E.G.	8.9	10 cm	Companion NGC 4647 glimpsed with 25 cm
M61	80	12 21	04 36	S.G.	10.0	15 cm	Visually a pale, indistinct smudge
M63	81	13 15	42 08	S.G.	9.5	6 cm	20 cm shows hint of structure
M64	82	12 56	21 48	S.G.	8.8	6 cm	Dark feature glimpsed with 15 cm
M65	83	11 18	13 13	S.G.	9.3	10 cm	Small aperture reveals spindle shape
M66	83	11 19	13 07	S.G.	8.4	10 cm	In field with M65; similar
M71	85	19 53	18 44	G.C.	9.0	10 cm	Dim; needs 20 cm for resolution
M74	87	01 36	15 41	S.G.	10.2	15 cm	Difficult: needs very dark sky
M76	88	01 41	51 28	P.Neb.	12.2	15 cm	Elongation detected with 15 cm
M77	89	02 42	00 04	S.G.	8.9	10 cm	Spiral arms suggested in 20-cm view
M78	92	05 46	00 02	D.Neb.	8.0	10 cm	40 cm shows hint of dust lane
M81	93	09 54	69 09	S.G.	7.9	6 cm	Bright oval patch of light
M82	93	09 54	69 47	G.Irr.	8.8	10 cm	In same field with M81; small aperture shows spindle shape
M83	94	13 36	−29 46	S.G.	10.1	15 cm	Pale, diffuse, difficult
M84	95	12 24	13 00	E.G.	9.7	10 cm	Circular, featureless
M85	94	12 24	18 18	G.SO.	9.3	10 cm	Small, bright, featureless
M86	95	12 25	13 03	E.G.	9.3	10 cm	In field with M84
M87	96	12 30	12 30	E.G.	9.2	10 cm	Same appearance with most apertures
M88	97	12 31	14 32	S.G.	10.2	15 cm	Hint of elongation seen with 15 cm
M89	98	12 35	12 40	E.G.	9.5	10 cm	Appears as pale circular light-patch

Object	Page	RA h m	Dec ° ′	Class	Mag	Tel. Aperture	Comments
M90	99	12 36	13 16	S.G.	10.0	15 cm	Bright nucleus visible
M92	100	17 17	43 10	G.C.	6.1	6 cm	Superb globular, resolved with 15 cm
M94	101	12 50	41 14	S.G.	7.9	6 cm	Very easy galaxy for small telescopes
M96	103	10 46	11 56	S.G.	9.1	10 cm	Circular, featureless
M97	104	11 14	55 08	P.Neb.	12.0	15 cm	Very dim; hint of features with 40 cm
M98	105	12 13	15 01	S.G.	10.7	15 cm	Faint, elongated
M99	106	12 18	14 32	S.G.	10.1	15 cm	Circular, featureless
M100	107	12 22	15 56	S.G.	10.6	15 cm	Pale, visually unimpressive
M101	109	14 02	54 27	S.G.	9.6	15 cm	Large; needs low magnification for sky-contrast
M102	106	15 06	55 50	E.G.	10.8	15 cm	Elongated oval without details
M103	111	01 32	60 35	O.C.	7.4	6 cm	Easily resolved at 30 ×
M104	113	12 39	− 11 31	S.G.	8.7	10 cm	Dust lane glimpsed with 25 cm
M105	103	10 47	12 42	E.G.	9.2	15 cm	Core, periphery distinguished with 30 cm
M106	114	12 18	47 25	S.G.	8.6	10 cm	Easy; elongated shape well seen
M107	117	16 31	− 13 02	G.C.	9.2	10 cm	Pale globular, unresolved with small aperture
M108	117	11 11	55 47	S.G.	10.7	15 cm	Distinct E–W elongation
M109	118	11 57	53 29	S.G.	10.8	15 cm	Dust-features visible in 40-cm telescope
M110	119	00 39	41 35	E.G.	9.4	10 cm	In field of M31, at very low power
NGC 185	43	00 38	48 14	E.G.	11.7	20 cm	Oval and very pale, but mottled
NGC 246	90	00 45	− 12 09	P.Neb.	8.5	15 cm	More elusive than magnitude suggests
NGC 253	91	00 47	− 25 34	S.G.	7.0	10 cm	Large, detailed, impressive with 25 cm
NGC 278	44	00 49	47 18	S.G.	10.5	20 cm	Tiny, bright nucleus visible with 20 cm
NGC 404	42	01 07	35 27	G.S0	11.5	40 cm	Swamped by β Andromeda in smaller apertures
NGC 457	71	01 16	58 04	O.C.	7.5	6 cm	Bright, loose cluster
NGC 663	112	01 43	61 01	O.C.	7.1	6 cm	Like a nebulous patch with 6 cm; resolved with 10 cm
NGC 869	73	02 16	56 55	O.C.	4.4	5 cm	Bright, compact group at low magnification
NGC 884	73	02 19	56 53	O.C.	4.7	5 cm	In field with NGC 869 at low power
NGC 891	45	02 19	42 07	S.G.	10.5	20 cm	A dim, narrow sliver of light
NGC 1023	48	02 37	38 52	E.G.	10.0	20 cm	Seen as a featureless oval with 15 cm
NGC 1245	44	03 11	47 03	O.C.	9.0	10 cm	A misty patch of faint stars
NGC 1907		05 25	35 17	O.C.	9.9	15 cm	Like a smudge of nebulosity, S of M38
NGC 1931	54	05 28	34 13	D.Neb.	—	30 cm	Glimpsed with 20 cm

Object	Page	RA h m	Dec ° ′	Class	Mag	Tel. Aperture	Comments
NGC 1973 etc.	58	05 33	−04 48	D.Neb.	—	15 cm	Trace of nebulosity glimpsed with 10 cm
NGC 2158	49	06 04	24 06	O.C.	12.5	20 cm	Unresolved patch of stars SW of M35
NGC 2237	60	06 30	04 54	D.Neb.	—	—	Overlies cluster NGC 2244; visually elusive even with very large aperture
NGC 2244	60	06 30	04 54	O.C./Neb	6.0	10 cm	10 cm shows cluster well; nebula needs larger aperture
NGC 2245	66	06 33	10 11	D.Neb.	8.5	20 cm	Small, comet-like nebulosity
NGC 2247	66	06 34	10 20	D.Neb.	—	—	Unlikely to be observed visually
NGC 2261	61	06 36	08 46	D.Neb.	10.0	10 cm	Has been glimpsed with 8 cm
NGC 2264	62	06 38	09 56	O.C./Neb.	6.0	10 cm	Amateur telescopes show cluster only
NGC 2392	50	07 28	20 57	P.Neb.	8.3	10 cm	Needs 30 cm for view of outer ring
NGC 2903	84	09 30	21 44	S.G.	9.9	15 cm	Some detail glimpsed with 15–20 cm
NGC 3185	63	10 15	21 56	S.G.	12.5	25 cm	Cluster of faint galaxies, in one field
NGC 3187		10 15	22 05	S.G.	13.0	25 cm	
NGC 3190		10 15	22 09	S.G.	12.0	25 cm	
NGC 3193		10 16	22 09	E.G.	12.0	25 cm	
NGC 3628	84	11 18	13 53	S.G.	10.9	15 cm	Appears in field with M65/66 at low magnification
NGC 4217	114	12 15	47 14	S.G.	11.9	20 cm	Edgewise orientation easily detected
NGC 4301	80	12 20	04 50	S.G.	15.0	—	Unlikely to be seen visually
NGC 4388	95	12 23	12 56	S.G.	11.0	20 cm	Dim, narrow sliver of light
NGC 4394	94	12 23	18 05	S.G.	12.2	20 cm	Appears as a tiny smudge, adjacent to M85
NGC 4395	99	12 23	33 49	S.G.	11.4	20 cm	Difficult; very low surface-brightness
NGC 4402	95	12 24	13 24	S.G.	11.5	25 cm	Faint, difficult object
NGC 4435	95	12 25	13 21	E.G.	11.8	20 cm	In field with M86
NGC 4438	95	12 25	13 17	S.G.	11.0	20 cm	Forms close pair with NGC 4435
NGC 4449	115	12 26	44 22	G.Irr.	9.2	10 cm	Some structure seen even with 6 cm
NGC 4485	116	12 28	41 58	E.G.	12.5	25 cm	Forms pair with NGC 4490
NGC 4490	116	12 28	41 55	S.G.	9.7	10 cm	Small aperture shows elongated shape
NGC 4535	68	12 32	08 28	S.G.	10.7	20 cm	Pale; requires very dark, transparent sky
NGC 4559	98	12 34	+28 14	S.G.	10.6	20 cm	Large spiral of very low surface-brightness
NGC 4564	79	12 34	11 43	E.G.	12.1	20 cm	In low-power field with NGC 4567/4568
NGC 4565	108	12 34	26 16	S.G.	10.2	15 cm	Glimpsed as narrow streak with 8 cm
NGC 4567	79	12 34	11 32	S.G.	12.0	20 cm	Perceived visually as single hazy spot
NGC 4568		12 34	11 31	S.G.	12.0	20 cm	
NGC 4631	102	12 40	32 49	S.G.	9.3	10 cm	Some detail observable with 20 cm
NGC 4638	78	12 40	11 43	E.G.	12.2	40 cm	Tiny, star-like, in small aperture

Object	Page	RA h m	Dec ° '	Class	Mag	Tel. Aperture	Comments
NGC 4647	78	12 41	11 51	S.G.	12.0	30 cm	Visually, merges with M60
NGC 4656	74	12 42	32 26	G.Irr.	11.0	20 cm	Faint, rather difficult galaxy
NGC 5907	110	15 15	56 31	S.G.	11.3	15 cm	Nucleus visible with 30–40 cm
NGC 6543	112	17 59	66 38	P.Neb.	9.0	10 cm	Bright, easy planetary; like 7662
NGC 6781	19	19 16	06 26	P.Neb.	12.0	20 cm	Large, but very low surface-brightness
NGC 6804	39	19 29	09 07	P.Neb.	12.0	20 cm	A faint, asymmetrical glow
NGC 6946	38	20 34	59 58	S.G.	11.0	25 cm	Barely perceived with 20 cm
NGC 6960	36	20 44	30 32	D.Neb.	—	30 cm	Extremely subtle
NGC 6992	37	20 54	31 30	D.Neb.	—	15 cm	Can be glimpsed with 10 cm
NGC 7006	86	20 59	16 00	G.C.	11.5	30 cm	Very tiny
NGC 7009	20	21 01	− 11 34	P.Neb.	9.0	10 cm	20 cm shows E and W projections
NGC 7217	54	22 56	31 07	S.G.	11.0	25 cm	Featureless, circular object
NGC 7293	21	22 27	− 21 06	P.Neb.	6.5	15 cm	More challenging than magnitude suggests
NGC 7317 NGC 7318 NGC 7319 NGC 7320	26	22 33	33 42	G.Cl.	15.0	40 cm	Barely glimpsed with this aperture
NGC 7331	25	22 35	34 10	S.G.	9.7	15 cm	Can be located with 10 cm
NGC 7332	27	22 35	23 32	G.SO	12.0	30 cm	NGC 7339 may be seen in same field
NGC 7479	28	23 02	12 03	S.G.	11.5	20 cm	Central bar visible with 40 cm
NGC 7635	72	23 19	60 54	D.Neb.	8.5	40 cm	Pale fan of nebulosity
NGC 7662	46	23 23	42 12	P.Neb.	8.5	6 cm	Bright, easy object
NGC 7814	12	00 01	15 51	S.G.	12.0	20 cm	Appears as oblong streak
Abell 2151	23	16 05	17 40	G.Cl.	15.0	—	Unlikely to be seen visually
B 33	59	05 39	02 32	O.Neb.	—	––	Rarely seen visually
IC 405	55	05 13	34 16	D.Neb.	—	––	Nebulosity not seen visually
IC 5146	35	21 50	47 10	D.Neb.	—	30 cm	Star cluster seen, but not nebulosity

BIBLIOGRAPHY

Bishop, Roy L., ed. *Observers Handbook 1983*,
   Toronto, Royal Astronomical Society of Canada,
   1983. (Annual.)

Bok, Bart J. and Bok, Priscilla F. *The Milky Way*,
   4th ed. Cambridge (Mass.), Harvard Univ. Press,
   1974.

Burnham, Robert, Jr. *Burnham's Celestial Handbook*.
   New York, Dover, 1978

de Vaucouleurs, Gerard and Vaucouleurs,
   Antoinette. *Reference Catalogue of Bright Galaxies*.
   Austin, Univ. of Texas Press, 1964.

Field, George B., *et al. The Redshift Controversy*.
   Reading (Mass.), Benjamin, 1973.

Gaposchkin, Cecilia Payne. *Stars and Clusters*.
   Cambridge (Mass.), Harvard Univ. Press, 1979.

Hodge, Paul W. *Galaxies and Cosmology*. New York,
   McGraw-Hill, 1966.

Hogg, Helen Sawyer. *The Stars Belong to Everyone*.
   Toronto, Doubleday, 1976.

Jones, Kenneth Glyn. *Messier's Nebulae and Star
   Clusters*. New York, American Elsevier, 1969.

Jones, Kenneth Glyn, ed. *Webb Society Deep-Sky
   Observer's Handbook II: Planetary and Gaseous
   Nebulae*. Short Hills (NJ), Enslow, 1979.

Mallas, John H. and Kreimer, Evered. *The Messier
   Album*. Cambridge (Mass.), Sky Publishing, 1978;
   Cambridge (UK), Cambridge Univ. Press, 1979.

Sandage, Allan, ed. *The Hubble Atlas of Galaxies*.
   Washington, DC, Carnegie Inst., 1961.

Shapley, Harlow. *Galaxies*, revised. New York,
   Atheneum, 1967.

Van den Bergh, S. "The Optically Brightest
   Galaxies", *Royal Astronomical Society of Canada
   Observer's Handbook 1982*, p. 153.

Vehrenberg, Hans. *Atlas of Deep-Sky Splendours*.
   Cambridge (Mass.), Sky Publishing, 1971.

Webb, T.W. *Celestial Objects for Common Telescopes*,
   edited and revised by Margaret W. Mayall. New
   York, Dover, 1962.

ACKNOWLEDGEMENTS

We are grateful to numerous friends and associates
who have provided practical advice and moral
support during our preparation of *The Cambridge
Deep-Sky Album*.

Of the following individuals and groups
special mention must be made: George Ball has
spent countless hours assisting in the design and
construction of the cold cameras described in the
book. Optical wizard Leo Van derByl painstakingly
figured several mirrors used in the photographic
work. He also graciously provided the superb
hilltop site on which the Newton observatory is
located. Arthur Vanzetta, John Wells and Moody
Kalbfleisch have all, at various times, offered
invaluable mechanical and technical assistance. Al
Donnelly, who acquired the 40-cm mirror-blank
and then allowed himself to be parted from it,
made the photographic telescope possible. We owe
a debt of gratitude also to Colin Scarfe and Jeremy
Tatum of the University of Victoria's
Physics/Astronomy Department, to Chris Aikman
of the Dominion Astrophysical Observatory, to the
Victoria Centre of the Royal Astronomical Society
of Canada, to the MacPherson Library and to the
Greater Victoria Public Library.

Jack Newton
Philip Teece

INDEX

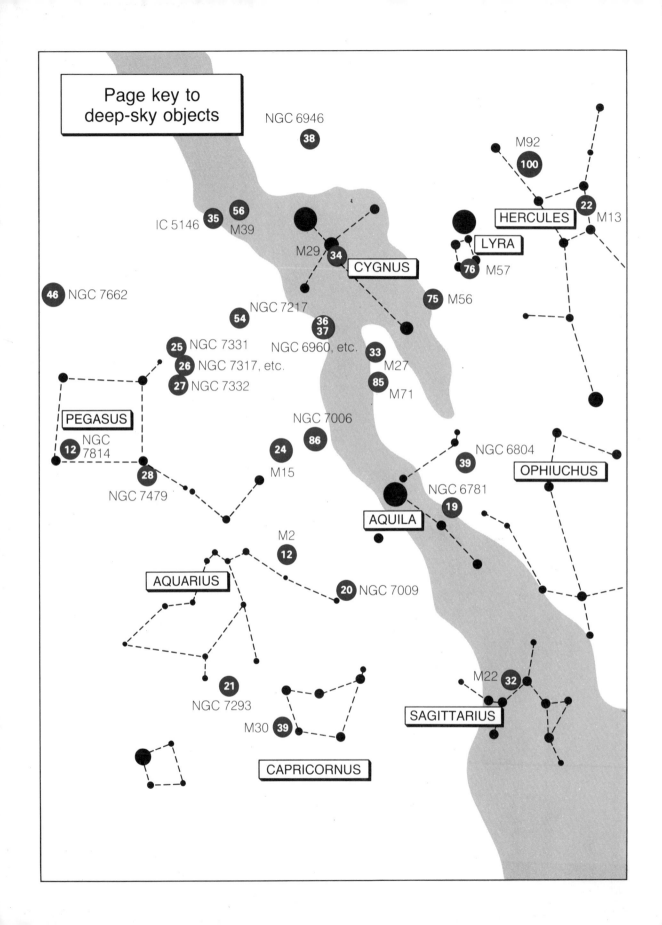